BUYER'S GUIDE ON CHOOSING THE RIGHT SEARCH ENGINE MARKETING AGENCIES & TOOLS

JEEV TRIKA

authorHOUSE®

AuthorHouse™
1663 Liberty Drive
Bloomington, IN 47403
www.authorhouse.com
Phone: 1-800-839-8640

First published by AuthorHouse 7/7/2010

ISBN: 978-1-4520-4465-1 (e)
ISBN: 978-1-4520-4463-7 (sc)
ISBN: 978-1-4520-4464-4 (hc)

Library of Congress Control Number: 2010909790

Printed in the United States of America
Bloomington, Indiana

This book is printed on acid-free paper.

CONTENTS

CHAPTER ONE:

EVALUATING WEB SERVICES AND SOFTWARE

Having an effective website can be your gateway to better market presence, more profits and an increased customer base. People use search engines each and every day to look for information and find solutions to their problems. Although it's somewhat difficult to nail down exact search engine use numbers, it's estimated that 49% of all Internet users engage a search engine each day in 2008. Compare that to 2002 when only about a third of Internet users visited a search engine and it becomes clear how important it is to have a an effective website. (Fallows)

Doing business online requires a website that works effectively 24/7. Unless you are already in the business in a field related to Internet marketing, you'll likely need a web services specialist or a specific piece of software that will guide you through the complex process of meeting the needs of search engines and visitors alike.

In this book, we'll go over the major categories of search engine optimization and web marketing that your business will need in order to have an effective presence online. Whether it's landing page optimization, content creation or PPC bid management software, you'll find the resources that you need in order to correctly evaluate service providers.

Web service providers come in all different price categories and levels of competency. When you hire a company to work on your website, whether it's working on the content of the website or the traffic coming in, you are handing them the keys to your company's future earnings. It makes sense to investigate their qualifications fully before investing your time and money in working with them. The same goes for any piece of software you may need to help manage your business. You need to understand what it should do for you so you can rely on it for the long term.

Not only is it good business sense to become familiar with the basics of the services you are going to buy but it also protects you from being scammed by the growing multitude of inexperienced providers who are hanging out their shingles. As search engine optimization and other web related services become more important to business, more and more fake experts are emerging in the field. A good looking website and claims of getting you to the first page of Google are not enough to prove the worth of a web services company. This book will arm you with the knowledge that you need to do your due diligence when you are locating and selecting a quality provider, service or software.

THE KEYS TO FINDING QUALITY PROVIDERS

Throughout the following chapters, we'll look in depth at many important areas of web services. However, there are many characteristics that are common across the board. Looking at these characteristics will help you see at a glance whether or not a company or software is worth your time and your money.

Fortunately, there is a very powerful tool at your disposal for evaluating SEO providers. Any provider worth his or her salt will have a website in place to engage potential clients and show off their skills. Their website will tell you at a glance about their website skills and can give you clues as to how they will approach your site. Even if their services don't relate directly to web design (i.e.: they are a social media optimization company) any provider in the web marketing field should have a good quality website. Make sure the website is updated and see if they have and maintain a blog. If they blog, there should be recent blog posts and lots of activity in the comment area. Keep in mind that a good looking website is not a guarantee of professionalism – but a poorly designed website is definitely a sign that the company doesn't have its act together.

If the website has all of the essential components (which you'll learn about in this book) and is ranking well for the terms you are searching for, then it's a safe bet that the company will be able to meet your needs. Look for a company that ranks within the first page of results for your search terms. For example, if you are looking for "landing page optimization company" they should be ranked high in the search engine results.

In addition to looking at their ranking and their website, you should view the company through the lens of its own specialty area. For example, if you are looking at a content creation company, read the copy on their website. Do a search for articles written by the owner and evaluate the copy. Would you be happy with content of this quality on your website? The same goes for any of the other areas. Look for a strong social media presence for a social media optimization company. Search for videos created by a video SEO firm and evaluate how well they are ranking and the quality of the content.

Another way to evaluate quality providers is to look for official validations and certifications. Depending on the industry, this may not be applicable, but several areas of SEO and web services have official certifications that the providers can test for and receive. Generally speaking, the providers who have taken the time to seek out and get the certification have more experience and knowledge.

There are also several industry wide rankings sites that point out which providers consistently give top quality service to their customers. Sites like topseos.com (http://www.topseos.com) evaluate and rank web service providers based on a number of factors. Look for designations from ranking sites on the provider sites. They can be good indications of the quality of service and the experience that a provider has.

Look for evidence on their website of a portfolio or a previous customer list. An experienced provider should have testimonials, samples or other signs that they have worked with others in the past. If they have a list of previous buyers, you can contact the buyers and inquire about the quality of their work.

Finally, ask around to get leads on quality providers from other website owners that you know. If you belong to any professional organizations or forums related to your industry, it's likely that many of the members have websites of their own and have had experiences with web marketing. Even if their experience was terrible, it will be helpful to know who they used so you can avoid the same fate.

General Red Flags

In this book, we'll highlight red flags that you should watch out for when selecting a SEO company or other type of web services provider. These are practices, strategies or lack thereof that should warn you that a company is not on the level. There are more specific suggestions in each chapter, but here are some general red flags to look out for.

- **Poor quality website:** Since you're interested in improving your own web presence, you know that a professional looking website is a sign of a stable and professional business. Since SEO companies and web service providers are working in this area, they need to have a top quality website that clearly defines their purpose and how they can help you.

- **Guarantees of placement or results:** As you'll learn in the following chapters, SEO and other areas of Internet marketing are evolving and constantly changing. Professional companies do not offer guarantees of getting you in the top spot of Google. The best a company can do for you is to use their knowledge and experience to improve upon your previous search engine, traffic, conversion or other type of results.

- **No clear contact information:** Lack of a contact phone number is a big red flag. Even though most of your communication will probably be done through email, professional companies are available via phone.

- **Underpriced or overpriced services:** By evaluating several companies, you should be able to get a feel for a general price range. Avoid any company that charges rock bottom rates. Their prices may be low for a reason. The same goes for overly overpriced services, unless your needs in that service area go above and beyond what the average service provider could give you.

- **Negative reviews on the Better Business Bureau or RipOffReport.com:** Both of these sites are great resources for investigating a company's legitimacy. The Better Business

Bureau ranks sites on a letter scale and RipOffReport.com collects reports from the general public on their experience with companies of all types. When it comes to the RipOffReport site, be sure to read through the comments because a company has the right to negate or counter the report generated by a former customer.

+ **No contract before work starts:** Professional companies spell out exactly what they are going to offer and how much you will be charged beforehand. Do not give out personal information, such as the login name and password to your hosting account, before having a contract with the company.

General Questions

Throughout this book, we'll go over specific questions that you should ask a company who will be providing you with services. However, there are some questions relate to the general nature of a web service company's business practices. Here are sample questions and what to look for in answers to those questions.

✓ **What does your pricing structure include and what other services do you offer?** They should offer a clear breakdown of what they will do, and what they won't. If they offer this in the form of a formal proposal, the proposal should be informative and not "sales-y."

✓ **How long have you been in business? How many clients have you helped?** Although their length of time in business is not always a factor in their professionalism, it doesn't hurt to know this. Generally, the longer a track record they have the more likely they'll be able to help your company. However, with some areas that are relatively new, like video SEO, even the experts have short track records with this type of service.

✓ **Can I speak with some of your clients?** A professional company won't be afraid to put you in touch with former clients. Since references and testimonials on site can be faked, this is always a good measure of protection.

✓ **How much work is expected on my part?** Your web services provider may have different ideas of what you will do and what they will do. Depending on your budget, your needs and your capabilities you may be taking steps with your website before the company gets involved or in the middle of the process. Have them spell out exactly what you'll be required to do so there is no confusion along the way.

✓ **What is the payment schedule?** This is good to know beforehand and will avoid any surprises with expected payments.

✓ **What type of reporting or results confirmation do you offer?** This will depend on the type of service you are purchasing, but there should be a monthly report or evaluation done on a consistent basis that shows the company's progress in the area you are hiring them to manage.

ABOUT THIS BOOK

This book will walk you through the basic best practices of each key area in Internet marketing. It will help you to understand what a service provider or software should be offering you in terms of service and help. With any other major purchase for your business, or your personal life, you'll likely do research and figure out what you need before you buy. The same is with Internet services. You need to know what to look for before you can make the right decision.

You can read this book straight through or jump to the section that interests you the most. Each section is a fully contained resource for that particular service area. When you need a new area of service for your website or Internet marketing development, you can refer to this book in order to find the right provider.

There are two basic categories of companies that you'll find in this book – service providers and software programs. There are slight variations in format between the two categories. For the services chapters, we've outlined the important areas of evaluation and included red flags and questions for each specific area. For the software chapters, we've outlined the key areas and then placed a list of red flags at the end of the chapter for your review. Since with a software purchase you'll likely be making your decision based on the website and perhaps a demo, you won't have a need for questions to ask.

It is our hope that this book become a relied upon resource that you can turn to time and time again as your web services needs grow and change. We have done our best to identify major areas of concern and give enough education on these topics so you can become an informed buyer. By reading through the book, learning the basics and paying close attention to the red flag areas, you can buy web services or software with confidence.

CHAPTER TWO:

SEARCH ENGINE
OPTIMIZATION COMPANIES

Search engines are the gateway between your customer and your website. If your website is optimized to appear for specific keywords related to your business, you'll have a better chance of being found by the people who will purchase your products or services. The goal of search engine optimization is to make it easier for your ideal customers to find you.

Of all areas of Internet marketing, search engine optimization is perhaps the most important. It's ironic that it is also the least understood by most online business owners and even some who claim to offer search engine optimization services. In this chapter, we hope to clear up some of the misconceptions to arm you with information to assist you in choosing a quality search engine optimization company.

Search engine optimization is an umbrella term for a series of steps and a set of techniques that will pull a website out of the vast chasm of the Internet and into the search engine rankings. SEO is vitally important to the success of a website because search itself is important.

Throughout this chapter, and the rest of the book, we'll be referring specifically to Google when it comes to search engines. Most companies will look exclusively at getting their clients ranked in Google because Google gets the most searches. Several search engines, like Ask.com for example, use the Google algorithm to aggregate their results.

First, we'll look over the nature of SEO and how that relates to your experience as a customer so you know what to expect from a company. You'll also learn some major red flags for working with a company and case studies of what it looks like when a company does things right.

SEARCH ENGINE OPTIMIZATION BASICS

Comprehensive search engine optimization strategy will help achieve two purposes – increasing traffic and increasing conversions. Getting traffic without getting conversions is like throwing spaghetti at a wall and seeing how much sticks. Conversely, having a well converting website with very little traffic is just as futile.

Typically, an SEO company will increase your rankings for particular keywords (based on research) by making changing to your site's existing content, adding in new content and, in some cases, restructuring the navigation of your website. All of these steps are done in order to rank for specific keyword phrases that have been determined by their research to be effective.

Over time, the search engine referrals to your site from target visitors will begin to increase. You need to keep in mind that SEO is a long term strategy for increasing traffic and not an overnight solution. The search engine rankings are based on a complex number of criteria that are used to sort and evaluate website content. While some components of the criteria are known (like the age of a site, the use of keywords and the links to the site), it is not known exactly how these factors are used or even what the factors are. In the case of Google, there is a specific algorithm used calculates the search engine rankings.

Even the best of SEO companies does not have the "keys to the kingdom", so to speak. Many components of the formula are known, but the specifics are kept secret by the search engines. The search engines use specific and private criteria in order to rank websites in a certain order. In addition to being secret, these criteria are also changed on a consistent basis in order to improve the usefulness of search engine results and prevent people from abusing the system. Google, Yahoo and other search engines will adopt slightly different rules for search engine ranking, sometimes on a weekly basis. In fact, the top ten rankings on search engines can change from day to day.

Bruce Clay, a well-recognized personality in the SEO field, states on his site "There are no search engine optimization secrets — just ranking and placement methodologies to follow in order to beat your competition in obtaining a high ranking for desired search keywords." (Clay) That's why it's so important to understand these methodologies so you know whether or not an SEO company is legitimate

SEO companies work within a specific framework for the content, the links and the structure of the site in order to optimize for keywords. However, there are many other components that figure into how well a website does in the search engine rankings.

If the rules for ranking sites changes, a website may drop from its previous search engine ranking. If a SEO firm that competes with your hired firm develops a new strategy that works better, the clients of that firm may be able to jump ahead of your site. If there's a new competitor in your niche, your sites may take a dip. If you change servers, you may also experience a jump in rankings.

Since most of these factors are out of the SEO companies' hands, achieving and maintaining a top ten ranking in just one of the search engines is very difficult. If it is attempted, it requires constant monitoring and keyword updating. SEO is an ongoing process so any company that promises specific results in a finite period of time doesn't have the methodology in place for effective search engine optimization.

A SEO company should stay on top of the methods of search engine optimization in order to be able to provide you with the best service possible. They should know about the latest trends in search engine optimization. By educating yourself on the basics of search engine optimization and the components of a successful campaign, you can tell how much your potential company has stayed on top of things.

General SEO Red Flags

+ **An overly simple SEO strategy:** Given the complex nature of search engine optimization, it's absolutely necessary that your SEO company provides you with a comprehensive strategy for increasing your search engine rankings and conversions. Due to budgetary or time constraints you may not choose to use the entire scope of their services, but they should at least offer a wide-ranging plan of services to choose from.

+ **A promise for a #1 ranking for any given keyword phrase:** Obtaining and holding a #1 ranking for a particular keyword is difficult, if not impossible, for most popular keyword phrases. Even the best of the best cannot give a guarantee as specific as "We'll boost your results to #1 within six weeks." It is just not possible to be that specific and give a deadline for a specific ranking. Any SEO firm that offer this type of guarantee is guaranteed to be one thing – a scam!

+ **Using outdated and ineffective methods for search engine optimization:** Since SEO is such a changing field with the rules and the competition frequently changing, your SEO company has to change with it. If they are using old and quasi-legal ways of obtaining search engine rankings, they are not only doing a bad job but they may be putting the future of your site in danger. The reason this type of guarantee cannot be met and should not be offered is because there is so much involved in SEO that is out of the SEO companies' hands.

+ **Free trial services:** Unscrupulous SEO companies can gain access to your site's data by promising a 30 day free trial in exchange for your password and access information.

+ **Flat rate services:** A true SEO professional will analyze your site and determine the level of service needed before giving you a quote for services. Watch out for providers that want to offer you flat rate services with no analysis or customization. Flat rate services are offered by inexperienced web designers or scammers.

+ **Unsolicited SEO offers:** Unscrupulous companies may send you a message claiming that your site isn't ranking for important keywords and try to sell you services (normally in the form of a flat rate packet). Professional SEO companies don't need to spam people in order to get clients. If the email seems like a form letter, avoid the company.

+ **Refusing to answer your questions:** SEO professionals all have a few tricks up their sleeves, but this doesn't mean they should leave you in the dark. Unscrupulous companies will try to talk above your head and keep you in the dark. Professional companies will listen to your questions and answer them. The basics of SEO are freely available online, so you're not paying them for their SEO knowledge. You're paying them for their experience.

General Questions to ask an SEO Company

✓ **What is your SEO process or methodology?** In this chapter, you'll learn what an SEO company should do to increase your search engine rankings and bring your site more targeted traffic. With the knowledge you gain in this chapter, you'll be able to evaluate their answer to this question. Their plan should include needs analysis, keyword research, on site optimization, offsite optimization and their reporting methods. We'll cover the details of each of these areas later in this chapter.

✓ **What kind of ranking guarantees can you provide?** The answer should be none.

✓ **What changes will you make to my website?** They should make changes based on what is most search engine friendly. This may include design changes, content changes and structure changes. Before any work begins, they should lay out their plan and explain the benefits of the plan.

✓ **What are your qualifications?** An SEO company should at least be able to provide a history of their company's efforts. Other qualifications to look for are Google Certification, Yahoo Search Master Ambassador and ranking on Topseos. com.

✓ **Do you participate and/or, are you a moderator for any of the SEO - Search Engine Optimization / SEM - Search Engine Marketing Forums? If so, what is your username and can you provide links to your most recent or notable discussions?** This may not be the case, but if they are it can be helpful in verifying their credibility. When the company responds with their information, you can do your due diligence.

✓ **How many search engine optimization campaigns have you been involved with? What was your role for those projects? How many are still active? How many are inactive? If inactive, why?** This will give you an idea of their experience and their time available to work on your project. It will also tell you something about how long clients work with them. Look for a company that has a good ratio of active to inactive. They should have solid reasons why accounts have become inactive.

✓ **Can you provide references and case studies (including your clients' ROI)?** You deserve to know how the SEO company has performed in the past before you spend money with them. A professional SEO company will be forthcoming with several examples of their performance.

✓ **Can you assure us that the optimization strategies and methods that you are utilizing fall under the criteria of Best Practices for the SEO/SEM Industry? Can we assume that this means no penalties for our website?** This is essential to establish with the SEO company early on. This lets them know that you are aware that there is a difference between good SEO and bad SEO. It makes them aware that you know that there are illegal steps they can take in order to make your website appear at falsely higher rankings.

✓ **How much work will be done on a permanent basis, i.e., if we cease doing business together this work and its effect on our rankings will remain in place?** This is another way of telling what kind of plans they have for the site. Genuine inward linking strategies will stand up even after you've terminated your contract with the company. On site optimization, content selection, website structure changes and inbound links will all hold up. Tactics that artificially boost your ranking and traffic will be discarded. This will tell you what portions of the site adjustments would be legally yours. Their answer should be that you would retain rights to all of it, indicating that they are using genuine methods.

✓ **What kind of relationship do you have with Google?** The answer should be none. An SEO company that claims a special relationship with Google is lying. There is Google certification available, but

✓ **How soon will I start seeing the results?** The answer should be "I don't know." The only thing your SEO company can do is optimize your site to the best of their abilities and with their expertise.

COMPONENTS OF AN EFFECTIVE SEO CAMPAIGN

Following is a selection of effective methods for SEO that will help increase your results and avoid penalization. Use these methods as criteria for determining whether or not an SEO company has your best interests in mind.

To better prepare you for working with an SEO company and determining whether or not that company is actually doing a good job, you need to learn about the methods for achieving good search engine rankings. Although the specifics of these activities are changed to adapt to new search engine rules, these methods are the building blocks by which you should judge any proposed SEO plan delivered by a company.

NEEDS ANALYSIS

An SEO company should be committed to helping you meet your needs, whatever they may be. The first step in working with an SEO company is an analysis of your needs in order to set goals for your work together. An SEO Campaign is like a battle plan. Without a strategic plan, you won't be able to win the battle. Since there are so many factors involved in achieving search engine ranking, an SEO company needs to have a strategy that reflects your goals.

Your interaction should begin with a comprehensive overview of your current web presence, your current results and your expected outcome from working with their company.

Some sample questions could include:

o   Are you tracking visitor conversions or other types of goals?

o   What is your average monthly spending on pay per click advertising?

o   Do you have an email subscriber list? If so, how many people are part of the list?

o   What are your objectives for this SEO campaign?

These questions and similar questions will help the SEO company determine what they are working with and what steps they need to take to help you achieve your goals. They will guide you toward defining your goals based on the answers that you give.

In addition to asking you these questions, they may perform a site diagnostic to analyze the technical data that is coming in from your website. This will include:

o   how many pages of your website are being indexed by Google

o   how many new visitors are coming to your site each month

o   what type of backlinks are coming to the site

o   how long visitors are spending on the site

o   what META tags are being used

o   what links are appearing on the pages

o   how the internal structure is set up

o   what content is being used and what graphics are being used

This comprehensive analysis will let them know exactly what they are working with and what might need to be changed when they start to optimize your site for the search engines. They may suggest implementing strategies you may not have thought about before.

Gap analysis techniques will also be involved. SEO gap analysis is basically an assessment of the deficiency in the site. It will take in the analysis of what is on the site and compare it to what needs to be on the site in order for your company to meet its traffic and conversion goals. They will analyze "the gap" between these two set of criteria (your existing status and your goals) and formulate a plan to help you bridge the gap.

By performing this analysis, they will begin to get a clearer picture of why you set out to find an SEO company and they will begin to formulate the battle plan to help you achieve your goal.

Case Study #1: YogaDirect

SEO Company: WPromote (http://www.wpromote.com)

Needs Analysis: YogaDirect is the largest direct importer and manufacturer of Yoga equipment in the U.S. While the brand was well established online, a large part of their organic traffic was from branded keywords searches. This meant that visitors who were looking in general for yoga accessories were not finding the site. The goal with YogaDirect was to increase their ranking for non-branded keywords. The challenge was that the site is in a crowded niche and YogaDirect's site was young compared to the competition.

Case Study #2: Green Acres Furniture

SEO Company: The Karcher Group (http://www.tkg.com/)

Needs Analysis: Green Acres Furniture is an Amish Country manufacturer that creates and sells custom Amish-built furniture. Since Amish furniture stores and very popular, Green Acres Furniture wanted to make their marketing distinct. Since most Amish manufacturers weren't using SEO, this was a natural choice. Their goal was to improve traffic and conversion on their existing website.

(source: http://www.tkg.com/seo-case-study-amish-furniture)

A quality SEO company will perform a comprehensive needs analysis before they get started. This helped them develop a foundation with which to build the rest of their SEO plans.

Red Flags for Needs Analysis

- ◆ **They tell you what your needs are:** A professional SEO company will spend time listening to your needs before determining which steps should be taken.

- ◆ **They give you a blanket approach to SEO:** Even if the SEO company listens to your needs, it's a red flag if they respond with a blanket approach to SEO. In this chapter, you'll learn about the customizations that your company should be making to your campaign.

- ◆ **They don't ask for traffic stats or previous conversion rates:** This data is essential for determining the benchmark from which your SEO company will begin their work.

Questions to Ask About Needs Analysis

- ✓ **How will you determine what my site's benchmarks will be?** They should determine the benchmarks based on your goals for your site, its current stats and the time of the contract.

- ✓ **How will you determine the gap between my site's current state and my goals?** They should use gap analysis techniques to compare your current benchmarks with your future goals.

- ✓ **What type of volume of traffic increase is reasonable to expect?** The SEO company should provide you with their best estimate and not a pie in the sky figure. Their response should be hype-free.

- ✓ **What techniques will you be using to achieve these results?** This will allow you to analyze their methodology and determine whether or not they are worth the money. They should be upfront with you and not evasive. If they hide under the "they are trade secrets" excuse, it's a bad sign.

KEYWORD ANALYSIS

Keywords are the first building block of effective SEO. Simply put, keywords are the terms that your customers use to find your site. When they are typing words into the search engine, those are keywords. Optimizing your pages for the right keywords will help your page show up where it will do the most good.

One of the goals of SEO is to increase traffic to your site but it has to be the right kind of traffic. 10,000 visitors a day to your gluten-free bakery website will do nothing for you unless those people are interested in buying your gluten-free bakery items. Targeted traffic can only be obtained with the proper use of keywords.

A good SEO company will continue their work with your site by doing keyword research. This research will include brainstorming for potential keywords, which is a process of writing down keywords that one would expect to drive traffic to your site. For example, if your business is golf course design, you'd expect your potential customers to find you using term like "golf course design" or "golf course development."

After brainstorming a list of keywords, the SEO company will generally look at which websites are currently ranking well for these keywords. You can test this out yourself by entering your main keywords into Google and viewing which websites are currently on the first page of the results.

Test this out in Yahoo, Bing and Ask as well to see which sites are consistently ranking well across the board. Your SEO company should do the same. By taking note of URLs that are using the keywords on the brainstorm list, an SEO company can take their keyword research to the next level. Since these URLs represent your competition, it's a good idea to know what other related terms they are ranked for.

SEO companies will look at the page source data for your competitors' pages in order to see what keywords they are ranking for. You can do the same simply by selecting "View" and then "Page Source" in your browser. To find the additional keywords, look for the "<META NAME="Keywords" CONTENT=" line. You can use this tactic to expand your keyword list and be sure that your SEO company is targeted the keywords they should be in order to compete with similar website.

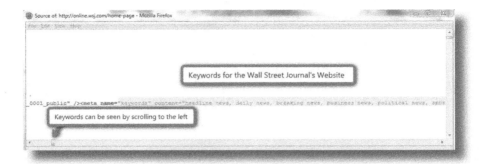

(source: http://online.wsj.com)

If you decide to look up these keywords yourself, it's important to note that you shouldn't copy the HTML code of your competitor's site. Simply look at the HTML to determine the keywords. If you copy the entire code and use it on your site, it will produce a duplicate listing in the search engines and it is considered to be a copyright violation. Your SEO company should follow the same practice. Keywords from other websites should be used purely as inspiration.

In addition to looking at competitor keywords, an SEO company will input the brainstormed keyword list into keyword search tools to determine other keywords that would be good to target. By using a keyword tool and recombining keywords to form keyword phrases, they should develop a large list of keywords.

You can test this out for yourself by using the free keyword research tool provided by Google at https://adwords.google.com/select/KeywordTool. This tool shows how often search terms are used by Google users and also gives suggestions for related words you or the SEO company may not have thought of.

Why all this focus on related keywords? It's important to understand that a quality SEO company will help you rank well among many different keyword terms. It's much more effective to have decent rankings among several related keywords than holding a number one spot with just one keyword. It's also much more practical this way since, as we already discussed, it is nearly impossible to create consistent a #1 ranking for major competitive terms.

A final step that a quality SEO company will use in order to develop keywords is to combine keywords into long tail phrases. "Long tail" keyword phrases are terms with four or more words that people will use to find a website. By combining keywords, your SEO company will be able to develop a long list of highly targeted keywords.

Long tail phrases are important to good search engine optimization because they are actually more effective in obtaining traffic that will lead to conversions. This is because they imply intent. Think about it this way. If you were in the market to buy custom cabinet knobs for your kitchen renovation, you may search in the following manner.

o   First, you'll just look up cabinet knobs to see what is out there. Then you may decide that you want brass cabinet knobs, which becomes your new search term.

o   After looking a while longer, you come realize that the look you are going for is really oil rubbed brass cabinet knobs in 1.5 inches. By typing in "oil rubbed brass cabinet knobs 1.5 inches" into the search engine, you'll get exactly what you are ready to buy.

As a retailer, it makes a whole lot more sense to have your page with oil rubbed brass cabinet knobs in 1.5 inches rank for that term rather than trying to get your entire site ranked for cabinet knobs. The searcher who is looking for that long tail keyword phrase is closer to making a buying decision, while people who are looking for cabinet knobs are more likely to be simply looking for information.

Multiple word keyword phrases will always convert better and will lead to increased sales. If an SEO company attempts to use a keyword targeting strategy that doesn't include long tail keywords, they aren't using keyword research effectively. With a little time and the Google Keywords Tool you should be able to come up with several possible long tail keyword phrases. An SEO company should be able to at least do the same if not a better job as you can.

Short, broad keyword phrases may seem like the best way to go from the layman's perspective. After all, if you are a company that sells gift bags, being ranked #1 for gift bags would be a big boon to your online sales. However, the practicality of that happening in a short period of time isn't very likely.

A good SEO company won't sell you pie in the sky ideas about your site being ranked well for a broad term. The truth is that cracking into the top 10 for a broad term like "gift bags" isn't practical. The websites that are in the top ten for a broad phrase like that have been in business for a while and have worked at developing the backlinks for that broad phrase.

A much better strategy is to focus on phrases in one of three categories:

+ Keywords that will realistically put your site in the top 10.

+ Keywords that your competition has missed.

+ Keywords that bring in highly targeted prospects.

When it comes to choosing how many phrases your site should be optimized for, it comes down to your SEO budget and the amount of web pages that you have. You can only realistically optimize a single page for one to two terms. This means that if you have a website with 5 to 10 pages, you're looking at 10 to 20 keywords tops.

Your SEO company should provide you with a long list of keywords to choose from. Some of them might not be appropriate for your market, but that's just because your SEO firm doesn't have the expertise that you do at the specific subject. They are looking to you to provide feedback.

Since a long list of keywords means more money spent on SEO, some people mistakenly believe that an SEO company is trying to squeeze money out of them by suggesting a lot of keywords. The keywords they suggest are yours to pick and choose from. Don't be surprised as well if some of the terms they are suggesting seem to get less traffic that you'd like each month. The amount of traffic is important to look at, but it's more important to look at how the traffic will react once they reach your website. The conversion rate is more important, so long tail keywords should be used more frequently.

Case Study #1: YogaDirect

Keyword Analysis: In order to bring first time visitors, WPromote focused on optimizing the site for popular Yoga relate keywords. By completing comprehensive keyword research and a site audit, they determined the specific non-branded keywords to focus on branding YogaDirect as a niche authority.

Case Study #2: Green Acres Furniture

Keyword Analysis: Considering the fact that many Amish furniture stores do not complete SEO for their websites, the Karcher Group was able to focus on several popular but un-optimized terms to create a list to work with for Green Acres Furniture. In order to optimize the site, they focused on general terms for Amish furniture as well as more specialized terms for each type of furniture. The keywords were specialized for each page based on the type of furniture that was being sold there.

A quality SEO company will combine keyword phrases and come up with many different variants of your main keyword phrases based on brainstorming, competitor phrases and keyword research. Once all these steps are completed it's time to put those keywords to good use.

RED FLAGS FOR KEYWORD RESEARCH

+ **Too few long tail keywords:** Long tail keywords should make up the bulk of your targeted keyword list. Long tail keywords are more specific so they lead to more targeted traffic and, therefore, better converting results.

+ **Doesn't include keywords that are part of your competitor's site:** As long as the competitor is ranking higher than you in the search engine listings, your SEO company should be looking to them for keyword ideas.

+ **Wants you to pay for every keyword that they get you to the #1 position for:** This sounds like a good idea in theory, but bad SEO companies may abuse the agreement. For example, they may optimize your website for "golf course design in Boca Raton, FL" and get it to #1. However, this won't make a difference in your business if no one is looking for that keyword term. The #1 listing doesn't matter as much as the usefulness of the keyword phrases. *Questions to Ask About Keyword Research*

✓ **What is your keyword strategy for my site?** They should tell you their strategy for choosing the keywords for your website and how it will help your traffic and conversions.

✓ **How will these keywords be implemented on my site?** They should be used in on page and off page optimization strategies (which are covered in the next sections). They should not be aggressively targeted or it will get your site banned. Your SEO company should use a mixture of different keywords for best results.

ON PAGE OPTIMIZATION

Finding keywords is an important part of the process because they give your SEO company something to work with. The optimization part of SEO can't happen until they have terms to optimize for. Your keywords are like your target. An SEO company will create the right kind of arrow and bow, in the form of on page optimization and offsite optimization.

On page optimization includes all components directly related to your website. It includes the content on your website and the structure that holds together the content. There are several basic areas of on page optimization that are important to know about. It is critical that your SEO company work on your site to optimize your pages, rather than just increase your amount of traffic. Following are a few key areas that your SEO company will use in order to optimize your site. These on site optimization techniques should be part of a comprehensive SEO strategy.

OPTIMIZATION USING WEBSITE CONTENT

Content is any text or images that appear on your web pages. The content of your website can also include video and audio. However, when it comes to search engine rankings the text and the words you use to label your images are the most important.

Some SEO professionals consider website content to be the single most important factor in determining the search engine results for a website. It's so important that there are entire agencies devoted to helping improve the written content on a website. We'll go over methods for reviewing those agencies in later chapters.

During the keyword analysis process, an SEO company will determine what keyword phrases your site should be ranking for. The next step is to tweak the site's content and add new content that reflects those findings. Content is considered to be important from the search engine's perspective because it helps enhance the user's experience. Google's job is not to give you criteria to help you get your site to the first position. Its job is to give searchers the type of information they are looking for based on their entered search terms. Sites with quality content that include keywords are much more likely to rank higher than content that is poorly written and stuffed with keywords.

Google cares so much about the quality of content on websites that they hire quality raters who are paid to review sites and confirm or deny whether they meet quality standards. Raters are given a list of things to look for – like excessive use of keywords, too many ads, bad web design, etc. –in order to qualify the content.

For this reason, content is often the first place that SEO professionals will look to make significant changes to your site. These changes will take place on the page's content and may include adding more additional pages to your site. From the long list of keywords that an SEO company develops, they will choose the most relevant keyword phrases and use those for optimization purposes.

The main way to optimize with content is to use those selected keywords in the text of the website. Each major keyword will get its own page where that phrase will be used repeatedly in order to optimize that page.

For example, a website that connects virtual employees with corporations will likely have keyword phrases like "hire virtual staff", "virtual employees" and "find telecommuting jobs" among many others. An SEO company dealing with this website will either tweak existing pages to develop a page for "hire virtual staff", a page for "virtual employees" and a page for "find telecommuting jobs" or will create three new pages to reflect these keyword terms.

However, it's not just enough to have pages to reflect those keywords. The content has to be written in such a way to attract both humans and visitors. When people first began trying to manipulate search engine rankings, they discovered that using keywords on the page would increase rankings. They began using these keywords as much as possible on the page, including hiding them in the white background text.

In response, search engines changed the guidelines for ranking and banned the excessive use of keywords and hidden keywords. These are two tactics you should watch out for when dealing with an SEO company. Shady keyword techniques fall into a grey area of search engine marketing - they can be done, but the downsides are many. These techniques may work in the short term (a week or two), but your website will instantly be penalized and prevented from increasing in the ranking for several months.

Keyword stuffing is "practice of loading a webpage with keywords in an attempt to manipulate a site's ranking in Google's search results" according to Google's official webmaster guidelines ("Keyword Stuffing").

Here are two examples of text optimized for the keyword phrase "bass fishing lures"

Keyword stuffed:

In order to find the right **bass fishing lures***, you need to look at all the* **bass fishing lures** *available in the* **bass fishing lures** *section of your local bass fishing shop. If you are able to find* **bass fishing lures** *that meet your needs, you can add them to your* **bass fishing lures** *collection and have enough bass fishing lures.* **Bass fishing lures** *come in many different shapes and sizes so it's important to find the right* **bass fishing lures** *for the type of bass fishing that you will be doing.*

Correctly optimized:

In order to find the right **bass fishing lures***, you should first think about what kind of fishing you are going to be doing. There is a big difference between bass fishing in a lake vs. bass fishing in the water. The type of lure you choose will also be dictated by the size of fish that you are trying to catch. If you want to get an expert opinion on picking your bass fishing lure, you need to either meet with a professional in a fishing shop, or do research based on your preferred catch and the location where you'll be fishing.*

The difference between the two paragraphs is obvious, although we've highlighted the keyword phrase to make it more noticeable. The first paragraph used the keyword phrase seven times in the space of about 100 words, bringing the keyword density to 7%.

The second paragraph has a keyword density rating of around 1% which is in the ideal range. According to Google's standards and empirical evidence it's been found that a 1% to 2% range for your keywords is ideal.

Besides keyword stuffing, hidden keywords are another unethical method of creating rankings for a specific keyword or set of keywords. This method is achieved by hiding keywords within the body of the page.

Text can be hidden in a variety of ways. According to Google's Web Master Tools page, the most common ways are ("Hidden Text and Links"):

+ Using white text on a white background

+ Including text behind an image

+ Using CSS to hide text

+ Setting the font size to 0

In short, being sneaky with keywords in the body of the website is a surefire way to get penalized by Google. A good SEO company won't try to cut corners when it comes to creating content for your website.

The first option is to optimize your existing content for the selected keywords. In many cases, you may already be conveying important information through the text on your website. You might have content on your site that is helpful to users that needs to be tweaked to meet search engine standards.

The second option is to create from scratch content that meets the need of search engines and website users alike. It is recommended that there is at least 500 words of original content (content that isn't copied from another source. If you've got a shopping related site, you can get away with having 250 words on each page.

Case Study #1: YogaDirect

Content Optimization: Once WPromote identified the non-brand keywords that they needed to focus on, they populated the site with unique content using those keywords. Each category page includes original content using the keywords associated with that category.

Case Study #2: Green Acres Furniture

Content Optimization: The Karcher Group identified the target keywords for the website and integrated them into each page with one to two paragraphs of unique content.

Above all, your SEO company should look at your content first and either help you improve it or guide you toward doing so with specific guidelines. For more on content creation and optimization, see the content creation chapter of this book.

RED FLAGS FOR CONTENT OPTIMIZATION

+ **Over-optimization for keyword phrases:** This is the process of using the same keyword phrases again and again, not only in content but in other optimization strategies. The content created for your site should use a variety of main keywords and related keywords.

+ **Poorly written content:** Even if content is properly optimized it can still be unintelligible to human eyes. The content should be well written and not just geared toward search engines.

+ **Irrelevant content:** The content should serve the purpose of your website. For example, an article optimized for "chocolate candy" on a gourmet chocolate website should not be about the dangers of chocolate candy or how to eliminate chocolate candy from your website.

QUESTIONS TO ASK ABOUT CONTENT OPTIMIZATION

- ✓ **Do you require that I change content on my website?** The answer should be yes. A company that says no is employing shady tactics or isn't familiar enough with SEO to know how much content matters in optimization.

- ✓ **Will there be new content or will you optimize the content already on the website?** If they answer yes to the previous question but no to this one, they may be trying to retrofit your old content to fit new purposes. This works in some cases, but more often than not your existing content doesn't fit the keywords. This may result in keyword stuffed content and unhelpful pages.

- ✓ **Will you be providing the content or will I need to provide it?** Either way is fine, but this should be clear at the offset of the relationship so you can make arrangements for content creation if necessary.

- ✓ **Will your staff create the content or will it be outsourced? Will it be outsourced overseas?** If the SEO company will be providing the content, they should be clear about where it is coming from. For quality purposes, your content should be written by native English speakers.

- ✓ **Will you be adding additional content that targets additional keywords over time?** Your SEO company's initial efforts should concentrate on specific core keywords. Professional level SEO companies will see these keywords as "trial" keywords that will be tested out with content and then they will build upon the results and create additional supporting keyword pages in the future.

OPTIMIZATION USING SITE STRUCTURE

Beyond the content that is on your pages, the way that your site is structured is also important in how the search engines respond to it. Think of the content on your site as the articles in a newspaper. The structure is the way those articles are displayed on the page. Without the layout, the articles would be unreadable. Without the site structure your site will be unreadable to the search engines.

The structure of your site will determine how the search engine spiders will read your website, what type of information they will gather, which content is most important, how much code they much read, and much, much more. Structuring your website in order to appeal to the search engines and the visitors is something that a good SEO company will focus on from the start. There is a slight disadvantage in that when an SEO company gets a hold of your website, it already has an internal structure that they need to work with. The extent to which they will need to edit your structure will depend on a variety of different factors.

First, how was your website built? There are many different ways to build a website and sometimes website design programs add in extra code that will make your website hard to crawl. The code should be as clean as possible and your SEO company will go through your site and remove unnecessary code. If you are having your website entirely rebuilt by the SEO company, they will be able to use style sheets in order to tighten up the code. Style sheets will help keep the design uniform across many different pages and will reduce the amount of code used on those pages. For more on structuring your website to motivate users to take action, see the landing page optimization chapter of this book.

Once the site has been set up with a structure that is agreeable to users and search engine spiders alike, the focus turns to the basic optimization of your site using tags and specific keywords to tell the spiders what your site is about.

META DATA

The search engine spiders are there to do several different jobs. One of their most important jobs is to determine what kind of content your site includes so it knows how to categorize it. Without the assistance of META tags, the spiders will look solely at the content of the site. Although the content of your site should be optimized, it's also important to use those same keywords in your META tags.

META tags are not a magical solution to website optimization. Using them correctly, you can expect a bit more control over the way that the search engines see your website. However, it takes a comprehensive organization and optimization of your website in order to get the best results.

A SEO company will alter the META tags to fit your keywords. Some of these tags are not seen by human visitors unless they select "View" and "Page Source" as we discussed in the section regarding using competitor sites to find keyword ideas. The META tags are information that is inserted into the code of the website to label it with descriptive sentences and keywords.

Here's an example of what the META data information looks like in the "View Page Source" section of an Internet browser, so you can identify what it looks like in your own website or in others.

(source: http://www.yogadirect.com)

Keep in mind that each page of your website will have this META information. It's important that your SEO company creates unique META information for each separate page. A page on yoga mats in the YogaDirect page will be optimized differently than a page on yoga balls. By adjusting the META tag information, your SEO company can help you get the most out of each page element.

ALT TAGS

A SEO company can also adjust the ALT tags to reflect your optimization keywords. ALT tags are an alternative text description used for images. The ALT text is displayed before the image is loaded. If for some reason the image isn't loaded, the text will be displayed. ALT tags can only be used with images, but they are important to incorporate.

ALT tags also provides an additional level of dimensionality to the website. It also increases the page's keyword density score and makes your site more relevant for your targeted keywords. Your SEO company should add ALT tags to all of your pictures and especially those that are used as links to other sections. Search engines give special consideration to ALT tags that are used with links.

ALT tags are required for images, so your SEO company might as well make use of them. They add another level of keyword relevance which tags your pages as being related to specific keywords. The ALT tags are especially important for navigation purposes. Search engines can't understand images and the ALT tags are used to describe the content of the images for search engines. By adding keywords to the ALT tags, your SEO company can ensure that your page will be enriched.

HEADER TAGS

In addition, your images will show up for your keyword terms under image search engines like Google Images (http://www.google.com/images). Once your SEO company has optimized your page, check the page source to make sure the ALT tags were used wisely, especially for your header image. Your header image is the first thing that the search engine spiders will reach, so it's essential that it has a proper ALT tag.

H1 Tags, which is short for Heading 1 tags, are another important element of structural search engine optimization. By using your target keywords in H1 tags, your SEO company can add another element that will contribute to your rankings. Headings, such as H1, H2 and H3 tags, are pieces of HTML code that make words stand out on the page.

H1 tags are the most commonly used, but H2, H3 and H4 can be used as well. As logic dictates, the subsequent heading tags should have less emphasis. The heading formatting can be incorporated into the style sheet of your website so all new pages will have the same heading fonts and sizes.

They should never use the same tag twice on a single page, and they should only be used when appropriate. The reason being is that an H1 tag should be the title of the page. It doesn't make sense to have two titles on the page. If your SEO company needs to break up the content, use an H2 tag and if the content needs to be broken up further, use an H3 tag. Generally, the amount of content on the page will only require two to three heading tags.

Heading tags shouldn't be used just because your SEO company wants to emphasize a keyword phrase. Making up headings to try to fool search engine spiders looks silly from the visitors' perspective and "spammy" from the search engine perspective. To top it off if the heading tags are used too much, their effectiveness is watered down. Look for your SEO company to use your main keywords as H1 tags in a logical way.

SPECIAL TEXT

If you find that your content needs to be broken up in several different places with many subheadings, your SEO company should consider using special text. Special text includes bold font, italicized font and colors. These formatting choices can be used for optimization because search engines view them as more important than regular text. This page from YogaDirect. com uses H1 tags, H2 tags and special text to separate the sections.

Mat Bags Offer Protection And Usefulness

For Sturdy Straps, Full-Length Zippers And Breathability In Yoga Mat Bags

Looking for yoga mat bags? You've come to the right place.

Dedicated yogis everywhere depend on our affordable, high-quality products for excellent performance and durability. Here are some key criteria to consider when shopping for the right bag.

The basics

- Sturdy straps. This will make it easier to tote your mat.
- Full-length zippers. You don't want to have to fuss with storing or removing your mat.
- Breathability. You'll want a bag that promotes good air flow, especially if you sweat on your mat during yoga.

The perks

- Roominess. Do you need your bag to hold anything beyond the mat, such as blocks?
- Extra compartments. A smaller pocket for keys and a mobile phone are handy. This makes searching for them after practice much easier.
- Long life. If you own a high-performing mat that's expected to last for many years, it deserves a reliable bag to match.

(source: http://www.yogadirect.com/yoga_matbags.html)

33

Looking at the code, the phrase "Mat Bags Offer Protection and Usefulness" is an H1 tag. "For Sturdy Straps, Full Length Zippers and Breathability in Yoga Mat Bags" is an H2 tag. "The basics" and "The perks" use bold font. The bold is used to direct the reader and add extra emphasis to those points. These phrases aren't as important as the title and the subtitle, so they don't get an official heading tag.

On the home page, YogaDirect uses bold to emphasize their message.

> **Welcome to YogaDirect.com, the internet's leader in high quality Yoga Mats at factory direct prices. YogaDirect is the largest direct importer and manufacturer of quality Yoga equipment in the world. From mats, bags, blocks and balls to Thera-Bands, DVDs, Pilates equipment and gifts, we carry it all. With over 10 years of quality service, our customers confidently shop with us knowing they are getting the best available products at unmatched low prices.**

(source: http://www.yogadirect.com)

This statement includes a lot of important keywords that are relevant to the site. By making them bold they draw the attention of human readers as well as the search engine spiders. Special text should only be used if it makes sense for both of your audiences.

INLINE TEXT LINKS

The last type of text to consider for optimization purposes is inline text links. There should be links to relevant parts of your website on the side of top menu of your websites. But you can also link from within the text itself to other parts of the website. It gives your website visitor a quick and easy way to visit other parts of your website. It also gives optimization weight to the target page.

If we were using inline text links with the fictional chocolate company example, we might link from the page on chocolate bars to the page on gift baskets. The text would look something like this:

> "Our gourmet chocolate bars make wonderful additions to our customized chocolate gift baskets."

The link to chocolate gift baskets is useful to the website visitor and it also adds extra weight to the term "chocolate gift baskets." It's theorized that inline text links are given more weight than links that are alone. When a link appears within the context of the content, it is more useful to the user. Since search engines look to give high rankings to sites that are valuable to users, it would follow that inline links are weighted more importantly than solo text links.

By following all these tips for structural optimization, your SEO company will be creating the perfect environment for search engines to read and understand your website. A website set up like this will invite spiders to quickly and easily index the site, leading to a better chance of your site being ranked for your keyword terms.

Case Study #1: YogaDirect

Structure Optimization: W Promote added unique META tags and descriptions for each page in order to target long-tail and product-specific keywords.

Case Study #2: Green Acres Furniture

Structure Optimization: The Karcher Group added META content that was directly related to the targeted keywords and optimized each internal page for relevant keywords.

RED FLAGS FOR OPTIMIZATION USING SITE STRUCTURE

+ **Not optimizing META data:** This should be a given. Even if the majority of their optimization techniques are done offsite, this is easy and quick to change.

+ **Stuffing the META data with long lists of similar keywords:** Just like with content, META data can be stuffed and appear "spammy" to the search engine spiders.

- **Not using ALT tags:** This shows the SEO company is not using all of the optimization tools available.

- **Using the same META content for each page:** Your site should be optimized on a per page basis and not straight across the board.

QUESTIONS TO ASK ABOUT OPTIMIZATION USING SITE STRUCTURE

✓ **Who owns the rights to the META content?** This is important to establish because it's an important legal issue if you leave their services. If they have the right assigned to them, they can legally bar you from using it or completely strip your site. Some states have laws that prevent transfer of ownership under work for hire agreements unless you treat the worker as an employee (meaning providing benefits). In addition to getting answer to this question, you should have a contract that explicitly states that the company agrees to transfer ownership rights of the META data over to you.

✓ **What elements will you use in the page structure to optimize my website for keywords?** The SEO company should use META tags, header tags, ALT tags and inline text links.

OPTIMIZATION USING INTERNAL LINKING

Links are generally spoken of in terms of offsite links coming back to your site. These links are important and we will discuss them in later sections. When it comes to on page optimization, the terms and the structure of the linking between pages is just as important. Internal links are the easiest to optimize and the easiest to maintain. They should be part of any complete SEO strategy. They are a subset of optimization using sit structure, but it's so important that it deserves its own section. By using an optimized internal linking structure, your SEO company can insure that your website is read properly by the search engine spiders.

The primary purpose of links within your website is to make sure that users and search engine spiders can find all of the pages on your website.

As a secondary purpose, it helps build the relevancy of a particular page to your main keyword phrase. It can also help your website users find information more quickly. Anything that helps the user will help you with search engines.

The inline text links mentioned in the previous section are an example of an internal linking structure. These links add emphasis to those particular words, help visitors navigate to important sections of the site and also help the spiders understand the structure of the site.

Many website owners think that the primary goal of linking is to help users get through the site. They focus on developing a linking structure that makes sense to the user. For example, if an inexperienced designer were to create a site for World Class Chocolates it may look like this.

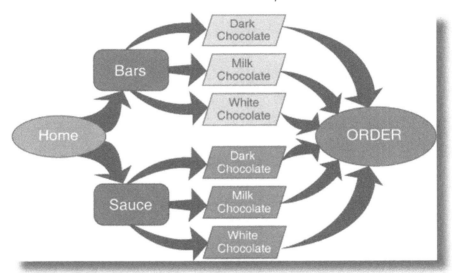

It makes logical sense to point users to the general categories first and then on to category specific specials, then products and then shipping and then payment. However, this means that the products themselves are two steps away from the home page. From a search engine's perspective, this means that the products aren't very important. Since they are really one of the most important things in terms of site usage and keywords, they should be given more strength.

Ideally, the website structure should be set up like this:

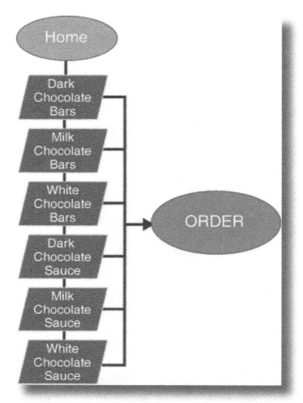

All pages are no more than two links away from the home page. If they can be directly linked from the home page that's even better. When a search engine spider "crawls" through your website, it indexes each of the pages and evaluates how each page relates to one another. However, the spiders don't know where it goes unless the structure of your site tells them where to go.

When your SEO company accurately directs the search engine spiders where to find all of your pages, it will increase your rankings. It will increase the number of pages that your website can rank for, which may mean that different part of your website show up for the same search term. This will increase your chances of getting traffic for that particular search term.

Internal linking is also helpful in maintaining the relevancy of a page to a keyword phase. When the search engine spiders crawl through your website, they are looking for indicators of what a page is about. A search engine spider crawls through the data, including the META data as discussed above, and by the time they are done they will have determined what a page is about. Part of this consideration is the incoming links and the keyword phase used in that link.

The vast majority of websites use keywords that are "hyper linked" to a URL rather than the raw URL itself. A raw URL is the information that appears in top of your internet browser, as seen below at the New York Times website.

(source: http://www.nytimes.com)

Links on websites rarely appear like this. They normally have a keyword phrase that lets the browser know what is on the other side of their click. This keyword phrase also identifies the link for the search engine spider. A keyword phrase link for our fictional World Class Chocolates example would look like this:

World Class Chocolates Home

In the website code, the link would appear as follows:

<a href=http://www.wordclasschocolates.com>World Class Chocolates Home</a>

Rather than just identify the home page with the word "Home" we'd use the company name. Not only does it help put the brand name into the search engine results but it also attaches the descriptive word "chocolates" to the home page. This tells the search engine spiders that the home page of our fictional site is related to chocolates and, more specifically, the brand name "World Class Chocolates." The text used to describe the link is called "the anchor text."

There are four very effective techniques to improving the internal linking structure, and we've already covered one in the last section – inline text links. The other three that your SEO company should use are text link navigation, footer links and sitemaps. All of these methods will create more keyword relevance and will produce a structure that the search engine spiders can understand.

TEXT LINK NAVIGATION

Text link navigation is recognizable on most websites in a column of links in the left hand side. This column is put on the left hand side for one purpose – the way that search engine spiders read pages. The search engine spiders start reading from left to right and by placing the column of internal links in the left hand side of the website those keyword rich text links are one of the first things the search engine spiders read. This means that your site is labeled with those keywords even before the spider reaches the rest of the page. These links will have a lot of weight to them, so it's important to choose the keywords wisely. Below is the optimized navigation menu from YogaDirect. Notice how each category is related to a keyword.

(source: http://www.yoga-direct.com)

As you can see, the product keywords are used as the anchor text for the menu links. A portion of the page source for the homepage shows how this was accomplished.

<a href="yoga-blocks-wholesale.html">Yoga Blocks</a></td>

Your anchor text should be based on your keywords, but your SEO company should not try to cram in the main phrase in every link.

Fortunately for this website, the main phrase "Yoga" is part of the other keyword phrases. However, this is not always the case. For example, think of how silly it would look to have an optimized menu that used the phrase "mail order chocolates" over and over again.

> Mail Order Chocolates Chocolate Bars
>
> Mail Order Chocolates Chocolate Sauce
>
> Mail Order Chocolates Chocolate Hard Candy

Not only does this look ridiculous to the website visitor, you could be marked as a spammer as well. In our fictional example, using "Chocolate Bars", "Chocolate Sauce" and "Chocolate Hard Candy" as anchor text for links is enough to identify our links and incorporate valuable keywords, even if they are not the term that we want to optimize the site for.

WPromote also used a linking technique with YogaDirect's site that helped increase keyword coverage and target longer tail keywords. In addition to establishing an internal linking structure using relevant keywords as anchor text, they also renamed the URLs of the internal pages.

For example, the phrase in the side bar "Yoga Blocks" leads to the URL:

http://www.yogadirect.com/yoga-blocks-wholesale.html

The phrase "Yoga Blankets" leads to the URL:

http://www.yogadirect.com/yoga-blankets-wholesale.html

Not all links are set up like this (with the word wholesale attached to the main keyword) but the fact that it is done in these two instances shows smart strategy on the part of WPromote. In this case, "yoga blocks wholesale" and "yoga blankets wholesale" may have less competition than the phrases "yoga blocks" and "yoga blankets." Alternatively, the longer tail keyword phrases may be more important to the optimization of the site.

No matter what the specific reason, the renaming of URLs adds another layer of optimization to the site. Your SEO company should utilize strategies like this to make the most of the data that the search engine spiders read. Using keywords in the anchor text and the URLs of text link navigation will further optimize the site.

Footers are an area of web design that can be either ignored or abused. They are helpful in getting spiders through the site but they shouldn't be used to list dozens of different keywords. This verges on spam and can result in each individual link losing more power. In some cases, unscrupulous SEO companies create footers that are larger than the content areas themselves. Footers should be kept clean and only include links to the most important pages within your site.

The Karcher Group made use of the footer section of the Green Acres website. In an older version of the site, the footer looked a lot like the footer used in stationary. It included the contact information for the company but no links.

skilled craftsmen in our own workshop facility. Great
care is taken to control the quality of each individual
piece of furniture.

Green Acres Furniture	Green Acres Furniture
2011 Route 19	7412 Massillon
Cranberry Township, PA 16066	Navarre, OH 44662

All images and content on this site © Green Acres Furniture

(source: http://web.archive.org/web/20070819065630/http://greenacresfurniture.com/)

By identifying important parts of the Green Acres Furniture website and adding text links in the footer section, the Karcher Group was able to add another level of optimization, as seen in this updated

(source: http://www.greenacresfurniture.com)

This improved header has links to the most important parts of the website – the furniture homepage, the room collections, the individual pieces and the about page. This footer is perfectly designed to lead website visitors and the search engine spiders to the right spots.

The final component of optimization using internal linking is the Sitemap. A Sitemap will show the search engine spiders and users all of the pages of your website in a hierarchical order. While the left hand navigation bar shows users some of the pages, the Sitemap catalogs everything on the site. This means that the entire site is accessible from one webpage, and it provides an "at a glance" view of the site for the website visitor, and more importantly, the search engine spiders.

Although Sitemaps provide navigation help, they are also very important for indexing purposes. This is particularly true if your website uses Flash or any other "special effects" in the navigation. Search engine spiders cannot index these special codes. They skip right over them, which can be a problem when it comes to rankings.

To circumvent this, SEO companies use a Sitemap to index the site and give search engines spiders some bearings. This means that web designers can be free to use techniques that make the site look spectacular without sacrificing optimization. The following example from the YogaDirect.com site.

(source: http://www.yogadirect.com/sitemap.html)

This Sitemap goes on for several page links and indexes all of the various parts of the website. There are approximately three times as many listings than there are in the left hand side navigation bar. This is the true representation of the content of YogaDirect and it helps the search engine spiders access all this information easily.

Your SEO company should implement a Sitemap on your website if it doesn't have one already. It doesn't need to be in a long list like the one used at YogaDirect. Some webmaster choose to make the site map part of the overall web design so it appears at the bottom of each page, kind of like an advanced footer.

Here's an example of the Sitemap/footer combination from Mozilla's Firefox (an Internet browser program).

(source: http://www.mozilla.com/en-US/)

There is no excuse not to use a Sitemap. Google has a free tool in its Webmaster Tools that allows you to quickly and easily create a Sitemap of your existing site. The benefits of having a Sitemap are too great to leave this out. For more details on the free tool you can visit:

http://www.google.com/support/webmasters/bin/answer.
py?hl=en&answer=34634.

These internal linking strategies will insure that website visitors and search engine spiders can find all of the components that make up your website. With proper use, they can increase the functionality of your website and help you get several pages indexed in Google for important keywords.

This wraps up the on-page optimization section and gives you the background that you need to be able to evaluate the on-page strategies proposed by an SEO company. However, before moving on to the offsite strategies, there is one more step that needs to be completed in order to speed up your ranking process.

Case Study #1: YogaDirect

Internal Linking Optimization: In order to improve YogaDirect's ranking, WPromote organized the URL structure to identify specific product categories and added product related keywords to the linking structure.

Case Study #2: *Green Acres Furniture*

Internal Linking Optimization: Because of the amount of products available on the site, Green Acres Furniture needed an organized site structure that allowed search engine spiders and humans to navigate through the sections.

RED FLAGS FOR INLINE TEXT LINKS

- **Using the same anchor text in the inline text links:** The goal should be diversity in keywords for site navigation and in text links.

- **No site map:** Sitemaps are an effective way to build strong internal links. Your SEO company should be using this method.

QUESTIONS TO ASK ABOUT INLINE TEXT LINKS

- ✓ **Do you use hidden links or hidden text?** These tactics will get your site penalized by Google and the other search engines.

- ✓ **Who owns the rights to the site map?** Much like META data, this is one area that unscrupulous SEO companies can use to bully their clients into staying with them.

SEARCH ENGINE SUBMISSIONS

Submitting your pages to search engines is something that should be done by your SEO company. However, there is a right way and a wrong way to do it. Search engine submission is essential because it speeds up the process of your website getting indexed by the major search engines.

Your SEO company should be submitting your website and significant pages manually. This means they visit the submission page for each search engine and type in the title of you page in order to submit it. There are several auto-submission tools and services that promise to submit your website to lots of different search engines and directories. On the surface, this may seem like a good idea. However, in the case of search engine submissions more doesn't necessarily mean better.

Submitting to Google, Yahoo and Bing (formerly MSN Search) is all your SEO company needs to do with regards to submitting to search engines. Beyond those three, there are diminished returns for submitting to a great number of search engines. Automated tools will submit your pages to a vast number of sites, but it won't be helpful to your rankings. It might even be harmful if you site obtains a link from a spam directory through automatic submission.

Here are the submission websites for the three major search engines:

o   Google - http://www.google.com/addurl/

o   Yahoo - https://siteexplorer.search.yahoo.com/submit

o   Bing - http://www.bing.com/docs/submit.aspx

In addition to your homepage, you should also submit the major pages of your website. For example, if we were submitting for World Class Chocolates, we'd submit the main page and the sitemap page, as well as each of the product pages. The shipping details page or about page would not be as important to submit. Any page that a visitor who is unfamiliar with your site could go to immediately and get value is a page that should be submitted to the search engines.

Submitting to search engines is not a guarantee that your site will be ranking. Many times submissions are discarded without being processed due to large numbers of people submitting low quality sites. Your SEO company should monitor the rankings of your site and resubmit if specific pages aren't showing up in the rankings.

It will take two to four weeks for your search engine submissions to be indexed. Your SEO company should keep track of the submissions and submit twice a month until your pages are indexed. Please keep in mind that there is a big difference between being indexed and being ranked.

Although you'll be paying for the time and energy that your SEO company uses to submit your website, search engine submission is free. Google, Yahoo and Bing all have free submission portals. However, there is also a paid option that is available if you want to guarantee being indexed by the Yahoo search engine.

Yahoo Paid Inclusion will quickly list your pages on the Yahoo Directory in exchange for the modest fee. This option is a good idea for commercial websites which can take a few months to get listed in Yahoo. Considering the fact that Yahoo reaches several hundred million unique users each month, it's a good idea to get in front of your potential customer quickly rather than waiting for it to happen with the free methods.

Your SEO company should set up a submission schedule based on how many pages you'd like to be indexed. They shouldn't submit more than two to three times per week and they should be paced out. There should be no more than five pages submitted per day, for best results.

You can check on the state of your submissions by typing the name of the site into the search engine in the following manner:

site: http://www.nameofyoursite.com

This will return the results of all the pages that are indexed for that site. Here's an example from Google using YogaDirect.com's website.

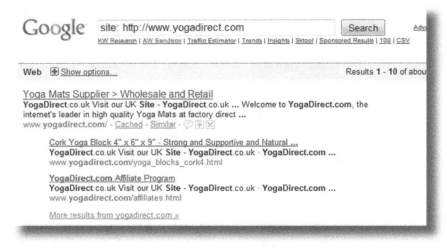

(source: Google search engine results for "site: http://www.yogadirect.com")

As you can see, several pages of the YogaDirect website are indexed. The main page, one of the category pages and the affiliates page are all showing up, and there are more results for this site as well. You can use this method to check up on the indexing process and make sure that your SEO company is properly submitting your sites to the search engines. In the next section, you'll learn how your company's website can move up the in the rankings with help from other websites.

RED FLAGS FOR SEARCH ENGINE SUBMISSIONS:

+ **Promises of getting a site indexed with 48 hours or less:** Your SEO company should focus on getting your site optimized before they even consider site submission. If they promise to submit your site, unseen, they don't have your best interests in mind.

+ **Promises of your site being submitted to thousands of search engines:** Professional SEO companies know that when it comes to submissions it's quality and not quantity that matters.

+ **Your site has not been indexed two to three months after it has been submitted.** This could be a sign that the SEO company did not submit as promised. It can also indicate that they implementing a submission strategy that got your website banned.

QUESTIONS TO ASK ABOUT SEARCH ENGINE SUBMISSIONS:

- ✓ **Where will my site be submitted?** The SEO company should submit to the major search engines only.

- ✓ **Will you submit to Yahoo Paid Inclusion?** If the answer is yes, ask for their cost. It should not be more than the Yahoo Paid Inclusion price plus the costs for the time involved. If the amount is too high, it might be more cost effective to submit to Yahoo Paid Inclusion directly.

- ✓ **Will you be using an autosubmitter or submitting the pages manually?** The pages should be submitted manually.

- ✓ **When will you submit my site?** The site should be submitted after all major on site optimization practices are completed.

- ✓ **How often will pages of my site be submitted?** Pages should be submitted every two weeks until all of the pages have been indexed.

- ✓ **Will you monitor the pages to see if they are being indexed?** The SEO company should keep track of what has been submitted and re-submit if necessary.

OFFSITE OPTIMIZATION

Up until this point, you've learned about the tools and techniques that an SEO company can use in order to optimize your website for particular keywords. The steps outlined will give you a streamlined and focused site that will appeal to visitors and search engine spiders alike.

In this section, we'll go over the importance of links in your search engine raking. These links, called "backlinks", are integral to increasing your website's exposure in the search engines. This is a basic overview of the link building process as there is an entire chapter devoted to evaluating the efforts of a link building company later in this book.

A big part of a search engine's ranking systems is the clout that a website has with other websites. A simplistic way to look at the process is that the site with the most "votes" for a keyword term will be ranked higher on the search engine listings. Links represent the basic relationship between websites and for the foreseeable future it's the best way that search engine spiders can determine the importance of a webpage. That's because links imply trust. If another website is linking to your site, it means that they trust you to provide the information related to the keyword they use as anchor text.

There are three main steps for a link building campaign are:

+ Determining existing inbound links.

+ Creating new inbound links.

+ Adding outbound links.

Case Study #2: Green Acres Furniture

The Karcher Group implemented a monthly strategy of adding quality backlinks to the Green Acres Furniture website and monitoring where links were coming from.

This section is a brief overview of the nature of links and how you can evaluate the link building process that your SEO company uses to help your website. In the link building chapter of this book we'll explore these issues in more detail as well as no follow links, finding appropriate outbound links, how to ask for backlinks and where to create backlinks using a few simple strategies.

RED FLAGS FOR OFFSITE OPTIMIZATION

+ **Promises of hundreds or thousands of backlinks:** Quality counts more than quantity. Your SEO company should build links that will help optimize your site and not just focus on a large amount.

- **Unclear about how the backlinks will be obtained:** The SEO company should share the general idea of their link building strategy.

- **No plan for outbound links:** Outbound links should be an essential part of their offsite optimization strategy.

- **Trying to create links without adjusting content:** Even if they aren't offering complete copywriting services, they should add some new content to your site or improve the existing content so that it is "link bait."

QUESTIONS TO ASK ABOUT OFFSITE OPTIMIZATION

- ✓ **What is your backlinking methodology?** They should tell you that backlinks are essential to helping optimize your site for certain terms. They should tell you that backlinking should include content preparation, obtaining legal backlinks and using outbound links effectively.

- ✓ **How will you obtain backlinks for my website?** Your SEO company should describe their process of obtaining legal backlinks through the methods shared later in this book.

- ✓ **How will you find sites to create outbound links?** They should search for related sites that will help build your optimization, and not just sites in your niche that have high page rank.

- ✓ **Do you use any black hat techniques to create backlinks?** If they say no to this question but mention using paid links, link farms or any other tactic that sounds sneaky they are leading you down the wrong path.

REPORTING METHODS

The final major component of an SEO company's relationship with you is their reporting methods. For the most part, SEO companies do work that their clients don't see firsthand, unless they know where to look.

In this chapter, you've learned how to identify on site optimization and off page optimization methods that your company may use. You've also learned what steps your SEO company should take at the offset of your relationship with regards to analyzing your current situation and making goals for improvement.

An SEO company's reporting methods are determining whether they are delivering what they've promised. Before you sign on the dotted line with a search engine company, they should make it clear to you how and when they will be reporting on their progress. In addition to delivering comprehensive reports, the SEO company should use those reports to suggest the next steps for your company. It's not just about delivering data in a timely manner but about delivering the right data and using that data to develop a plan for the future.

Most SEO companies offer a monthly update report which details what steps were taken in that month and the results that were achieved. They will have different ways of categorizing their data, but they might look something like this:

-Keyword Selection and Usage Report

-SEO overview

-Ranking Report

-On Page Optimization report

-Outbound Links report

-Traffic and backlinks report

Your SEO company should provide you with a complete breakdown of the steps that they took and depending on the type of work they are doing for you, the previous categories may be too many or too little. No matter what the level of work, they should detail their steps. These steps should be directly related to the goals that they established for your site during the initial site evaluation.

The SEO company can provide details on the steps that they take, but there are many different methods by which they can evaluate their results. Being familiar with these methods can help you understand how they are evaluating results and what the advantages and drawbacks to those methods may be. Here are the various methods of tracking SEO changes that SEO companies use.

o   *Checking Manually*

This may be the method that you'll use yourself to check on the status of the SEO company's progress. SEO companies can also use this method themselves because it's a way to visually confirm the results of SEO methods without violating any search engine rules about automated rank-checking software.

The main disadvantage is that it is very time consuming, which is why most SEO companies don't rely on it exclusively in order to report their SEO progress back to you. For example, if they are targeting 20 different phrases for your website and are looking at the first 10 pages of search results, leaves them sorting through 2000 pages to check your status.

o   *Rank checking software*

The main benefit of this method of evaluation is that it is very efficient. It saves a lot of time and is generally very accurate, as long as a quality software program is used. Just enter the keywords that your site is being targeted for and the search engines you'd like to check on. You also need to add in how deeply you want the software to look in the search engines.

The main disadvantage to this method is that it violates the search engine rules about using their system and running a large number of searches at once. However, by inserting Google's API into the software you can run 1000 automated queries per day without running the risk of violating the search engine terms.

There are several software programs that will do automatic rank checking, including Advanced Web Ranking, Agent Web Ranking and Rank Tracker. They range anywhere from $150 to $220 for a one year license.

o   *Online services*

These run similar checks to what you can get with software, but they don't have the high costs associated with a software license. A few examples are Free Monitor for Google, Rank Checker from SEOBook.com and Parameter.

These services have a good price tag (free) but still take up a lot more time than automated software. However, they may be the perfect fit for your personal SEO checking. Let the SEO company pay for the software and just double check their results with an online service.

In addition to checking the search engine placement for various keyword phrases your SEO company should also be monitoring links in order to report back to you. Placing links and obtaining backlinks simply isn't enough. They did an audit of your links to begin with and now it is time to see how those links have grown.

Links should be checked on a weekly basis and your SEO company should also make it a habit of checking on the links of your main competitors. If you want your site to keep an edge, it has to remain competitive. Just because your number one competitor had 1,000 backlinks at the beginning of May doesn't mean that there will be the same number of backlinks at the beginning of June. They are spending time developing backlinks just as you are.

There are automated tools that your SEO company can use to monitor links and they will report to you on your link growth each month, if it is in your contract to do so. You can also monitor your own links manually using the method described in the backlinks section of this chapter.

Automated tools, however, can only do half of the work that an SEO company should do with reporting. Look for a company that can communicate intelligently about the results and can formulate a plan based on the reports. The monthly reporting should include a hand-created summary that will interpret the data for you and help your company work toward

RED FLAGS FOR REPORTING METHODS

+ **No reporting or expecting you to track the results yourself:** No matter how few services your SEO company assists you with, they should always report their results.

+ **Not checking on competitor links and ranking:** Monitoring the competition should be an ongoing part of your SEO company's services.

+ **Not relating reporting methods to your business goals:** In the needs analysis step, the SEO company should have established goals for your business. The reports should directly relate to those goals and show how those goals are being met. The company should add their own analysis to the automated reports in order to interpret the results and plan for your future.

QUESTIONS TO ASK ABOUT REPORTING METHODS

✓ **How will you be tracking the progress of my site?** They should indicate whether they will be manually tracking your progress or using specific software to test your results.

✓ **How often will I be updated on the progress of my site?** They should update you on a monthly basis. They should communicate with you during that month about following through on their goals for your site based on the previous month's result.

RESULTS OF SEO CASE STUDIES

Throughout this chapter, we've tracked the SEO efforts of two companies, WPromote with YogaDirect and The Karcher Group with Green Acres Furniture. They have used the five areas of search engine optimization: needs analysis, keyword analysis, on page optimization, off page optimization and reporting methods. Where appropriate, we've included examples from their actual SEO processes to not only show you what methods should be used but to give you real life examples of how it is done. However, all these examples would be useless without proof of the results that were achieved.

WPROMOTE'S CASE STUDY FOR YOGADIRECT

After three months of work with WPromote, YogaDirect experienced a traffic increase of 28%. WPromote tracked the competition during the same period of time and noted that the top two competitors only had an increase in traffic of 22.10% and 19.68% during the same period of time. WPromote also compared YogaDirect's traffic during their service period to similar results the previous year. Since the last year, YogaDirect saw a 40.31% increase in overall traffic. The main competitor saw a 40.86% annual decrease in the same period of time.

WPromote was able to create a #1 ranking for three keywords in three different search engines. The site was listed #1 for "yoga mat" in Bing with 2,680,000 competing results. They ranked "How to use yoga blocks" in Google with 7,020,000 competing results. YogaDirect was first for "yoga balls" in Yahoo with 8,010,00 competing results.

Other significant gains included a +28 ranking increase for "yoga equipment" in Google with 4,500,000 competing results. "Yoga mats" saw a +10 ranking increase in Google with 2,930,000 competing results. "Yoga equipment" increased 27 rankings with 36,000,000 repeating results in Yahoo.

Finally, WPromote was able to obtain new rankings for several keywords. YogaDirect ranked #2 for "Pilates rings" in Bing with 2,940,000 competing results. "Yoga blocks" achieved a #2 ranking among 6,340,000 results in Yahoo. YogaDirect ranked #4 for "yoga mat rolls" with 201,000 competing results in Google.

THE KARCHER GROUP'S CASE STUDY FOR GREEN ACRES FURNITURE

The main metric of improvement for Green Acres Furniture was their link popularity. Their total number of links from other sites increased by 193%. Their overall unique visitors increased by 55%. Conversions increased by 230%. In addition to those results, Green Acres furniture saw a 132%

increase in search engine referrals and a 30% increase in pages indexed by Google.

CONCLUSION

This chapter represents the foundation of what you should look for in a general service SEO company. The guidelines in this chapter can be used to identify and test a professional SEO company's services. Use them as a cheat sheet for cutting through the unscrupulous SEO companies out there (and there are many!) to get the help that you really need from qualified professionals. SEO is constantly evolving and a quality provider should show proficiency in these basics as well as the ability to adapt to the changes in the industry.

CHAPTER THREE:

PAY PER CLICK MANAGEMENT COMPANIES

Pay per click advertising, like search engine optimization, is one of the areas of search engine marketing that is essential and important. With pay per click ads you can place your company and your message directly in front of web searchers who need and want your services. With the right keyword choices, ad creation and bidding strategy your advertisement can be in front of your target market within hours.

As more companies in your niche become involved in pay per click marketing, the competition will become more intense. Developing a PPC campaign that is effective without wasting advertising dollars will become a more demanding activity. With the guidance of an experienced PPC company, you can optimize your PPC costs and target keywords that will bring in new leads.

Pay per click is one of the most competitive advertising models on the web. Since there is such a large learning curve to pay per click it is in your best interest to find and hire a quality pay per click agency to help manage your advertising. In this chapter, we'll cover the benefits of using PPC advertising and a broad overview of pay per click strategy. Then we'll look at several factors to help you determine whether or not a PPC company is taking the right steps in their advertising strategy for your website.

The Nature of Pay Per Click and Web Searcher Behavior

Pay per click ads are sponsored links that appear when you search for a specific term. You may see them above the regular search engine listings or to the side. Here is an example of the Google paid advertising for the term "pay per click advertising."

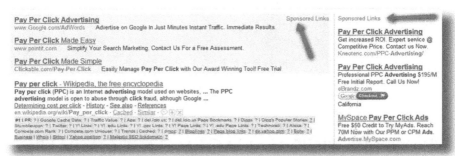

(source: http://www.google.com/search?hl=en&q=pay+per+click&sourceid=navclient-ff&rlz=1B3GGGL_enUS357US358&ie=UTF-8)

As shown in the example, the advertisements are listed in a specific order. The rank of the ads is determined by an open bidding system. Advertisers will place bids based on the amount that they are willing to pay when someone clicks on their ad. There is a cost per click, which gives the advertising model its name. The terms "pay per click" and "cost per click" are often interchanged.

Within the bidding system, the user that bids the highest cost per click will be shown at the top of the ad block. However, some search engines like Google have made this a little more difficult by incorporating other factors that determine which ads show up first. With the right bid amount, you can have your ad on the first page of Google within minutes, depending on your keyword choices and your budget. Considering the fact that only 28% of searchers click within the second or third pages of search results, it pays to be on the first page. PPC advertising is the fastest way to do it. (Elliance)

Pay per click ads are normally labeled as Sponsored Links, as they are in the previous example. They may also be referred to as Paid Links, Sponsored Listings or Featured Listings. With the tendency for people to avoid being "sold to", logic would dictate that advertisements would be shunned in online search results. However, according to the Pew Internet and American Life study conducted by Deborah Fallows, 62% of searchers were unaware of the distinction between paid and unpaid results on the search engine results page. Only 18% of searchers overall said that they could tell which results are paid and sponsored and which are not, even with the labels in place. (Fallows)

What does this mean for someone who wants to use pay per click advertising? It's a green light for using pay per click advertising. Other results from the same survey showed that:

+ 70% of searchers are okay with the concept of paid or sponsored results.

+ 74% of internet users under 30 years say it is fine for search engines to offer paid and unpaid results, compared to 71% of those 30 – 49 years, 67% of those 50 – 64 years and 60% of those over 65 years.

This is clearly an indicator that paid search can be as effective as natural search engine listings. SEO can get you less expensive leads in the long run but if you're looking to get an instant impact from your advertising, you can't go wrong with PPC.

MAJOR PLAYERS IN PAY PER CLICK MARKETING

Unlike SEO, where you optimize your website for all the search engines, your PPC marketing has to be aimed at one search engine. These search engines host pay per click ads in the margins of their organic search engine results.

Google accounts for approximately 81.2% of the pay per click market share, so the Google Adwords pay per click program is a natural choice for PPC. Although most websites aim to rank in the organic Google results, having Google Adwords placement can increase your site's exposure. Google has a minimum bid of $0.01 per click but normally bid rates are much higher than that. Ads are ranked in the advertising blog based on their bid amount with the higher bid showing up first, but Google also has additional factors that will determine ranking which are secrets of the Google algorithm. Despite its popularity, according to AdGooRoo, Google has the lowest conversion rates of the major pay per click search engines. However, it has so much more volume and a decent ROI that it makes the lower conversions well worth it. (AdGooRoo)

The other two major pay per click search engines are Yahoo and Bing. Between Google, Yahoo and Bing your PPC company can put together a comprehensive and successful PPC marketing campaign. Most professional PPC companies do not spend money or time dealing with other pay per click search engines.

Yahoo offers less expensive clicks but at a lower search volume. It accounts for 26% of all PPC purchases. Yahoo has increased conversions which make up for the lower search volume. Bing, formerly MSN, has the lowest volume for all but a few categories. It has 12.6% of the pay per click market share. However, according to AdGooRoo's third quarter report for search engine marketing in 2009, it has the highest conversion rates of all three. (AdGooRoo)

Your pay per click company should focus its efforts on advertising on these three search engines. Even though the cost of advertising may be higher than it is with second tier PPC search engines, it's well worth it to advertise where the most people are conducting their searches.

TYPES OF PAY PER CLICK PROVIDERS

In the pay per click industry, there are several different categories of providers that you can choose from. Understanding the basic types, including their advantages and weaknesses can assist you with making the right choice.

The first type is the solo PPC expert. This person is an independent expert who will be handling all of your needs. Unlike large PPC companies, the person who you are speaking with will be the person who will handle your account. Their fees are normally reasonable because they don't have the overhead of large companies. You may be able to get a PPC expert for a very low cost, depending on their level of expertise. The only drawback is because they are working independently they may not be able to handle a large volume of work.

Another type of PPC company is a full service PPC firm. This company has a large staff and generally a long track record to justify their high prices. They are very stable and reliable because they support a large number of staff. They generally have more sophisticated practices than the other options. The major downside is that they generally have the highest prices.

PPC management services often come as a service packaged with other search engine services and web design services. For example, a web development firm may offer PPC management or marketing companies may add on online advertising to their services. For the most part, these arrangements never work well. PPC is so intricate that it requires specialized knowledge in order to offer adequate service. Knowledge about another part of internet services does not equal knowledge about PPC.

PPC MANAGEMENT OVERVIEW

A pay per click company can help you overcome the learning curve with paid advertising. PPC is very intricate at every step of the process. There is a specific system to creating ads that convert and bidding on keywords so that targeted traffic will get to your site. If you are trying to learn pay per click yourself, it could mean months and of study and thousands of dollars lost in order to master the process.

There are three basic areas of pay per click management. These categories can be further broken down into additional aspects of PPC management:

+ Keyword selection

+ Ad copy

+ Landing pages

The keyword selection will determine what keywords your ad will show up for. Ideally, your advertising keywords will be directly related to your offer and targeted to be presented in front of a profitable segment of your market. For example, targeting ads on the keyword phrase "information on truck parts" is not as effective as targeting ads on the keyword phrase "chrome projector headlights with halo."

Your ad copy is limited to four lines of text: the headline, two lines of copy and a link to your website. In the space of just a few characters, you'll have to grab the attention of the search engine user and provide a call to action that will inspire them to click on your ad.

Finally, there's the landing page. All the paid traffic in the world won't matter if your offer isn't attractive to your target market. Although some PPC companies don't focus on making your landing page optimized for your offer, it should definitely be a factor in their management of your site. The details on landing page optimization will be covered in a later chapter in this book.

Each pay per click search engine has its own process. In order to do it yourself you have to learn how the systems work. You have to learn how to target the ads, write the ad copy, learn how to bid, where to rank and what terms to pick. In addition, you will be competing against advertisers that are far more experienced. Their search engine campaigns may have been in effect for months.

Rather than spend your time and money on mastering the process, you can allow a PPC company to do the heavy lifting for you. You may be able to pick up the basics of pay per click in a few days but mastering it takes years.

These aspects can be analyzed in the components in the following sections. These components will help you evaluate whether or not a pay per click company is on the right track with your PPC management. Your PPC company should stay on top of changes in the pay per click industry. By educating yourself on the basics of PPC so you know the process, you'll be able to tell if your PPC company is following the correct methodology.

Choosing a PPC company that understands your marketing objectives is essential to spending you advertising money wisely. If the company is not qualified or isn't in line with your objectives, you could end up with a poor click through rate, low quality scores for your ads, untargeted keyword lists and overzealous bidding which can cost you money. In addition to these qualities, you can rely on the rankings at topseos.com, which hosts a listing of the current top PPC companies online.

GENERAL PPC RED FLAGS

These are general guidelines to look out for when you are researching a PPC company. There are many types of PPC companies or PPC practices that should be avoided to protect your business and protect your wallet. In this chapter, we'll also go over additional red flags that are specific to individual aspects of PPC management.

- **Offering free trials:** A free consultation is one thing but a free trial for PPC services should be avoided. In some cases, unscrupulous businesses will offer free trials in order to gain access to your information.

- **Prices that are too low:** Comparing a prospective PPC service to others in the field, you should be able to determine the average price for the breadth of work that you are looking to receive. Really low cost companies are probably outsourcing your PPC management overseas.

- **Requiring that you to pay them first and then they'll pay the PPC search engines:** PPC agencies that operate this way pitch this service as being helpful to you. They claim that this arrangement will be stress free on your part. However, the drawback is that you never see your PPC accounts. There is no accountability for the PPC company because you won't be able to control how your money is being spent. Without access to your accounts, you won't have any of your conversion data if you decide to leave their services.

- **Flat rate services:** This is one of the biggest red flags for scamming. Pay per click bidding takes place in real time and should be customized to each specific campaign and company. There is no possible way that PPC management can be completed effectively with one flat rate. Monthly flat rates are pretty common, but a one-time fee flat rate isn't professional level service.

- **Claiming an "insider relationship" with Google:** There is a Google certification program for pay per click advertising (which we'll briefly discuss in the next section). But claiming an "insider relationship" is something completely different and completely impossible. Another common claim related to Google is that a company gets "extra discounts" because they buy so much advertising. This is also impossible. Google does not offer lower click rates for different companies.

- **Using lower ranked pay per click search engines:** Your PPC management company should concentrate on Google, Yahoo and Bing exclusively. The ROI for these search engines doesn't justify the time or money spent.

- **Unclear budget:** Your PPC company should make it clear what portion of the money you are paying is being used on advertising and which portion is being used as their service fees. This should be spelled out clearly in a contract. Do NOT trust the company to determine how the money should be spent. It should be clarified before you pay anything.

+ **Unsolicited offers for help:** You may occasionally get email messages offering PPC services. You should seek out PPC companies and not the other way around.

+ **An advertising agency that acts as a PPC agency:** Many traditional advertising companies are making their way into online advertising as a response to the changing field. However, online advertising and traditional advertising are two different creatures. Just because an advertising agency has been around for two decades doesn't mean they are the best choice for your PPC needs. Don't be seduced by a long advertising track record without proof that their experience directly relates to PPC.

+ **Relying solely automated pay per click management:** There are several different PPC management software programs available on the market. Most PPC managers use at least one in order to help inform their results. However, there is an art to PPC that cannot be replaced by software alone. Any PPC company that just tries to sell you the software or is unclear about the "human element" in their PPC company should be avoided.

+ **A 100% success record:** This may seem like it's a good thing, but all major PPC companies have had failures in their past. PPC is not absolute and there may be times where a company's objectives don't fit the advertising model. Experienced PPC firms will have had enough clients to have a few failures in their history. If a company claims a 100% success record they are either lying or inexperienced.

+ **Pressure to sign a contract before you've been able to investigate their methods:** Although there is a bit of secrecy when it comes to PPC methods, you should be able to determine how your PPC company will handle your account once they earn your business. You should be presented with a detailed accounting of the risk vs. reward for your PPC efforts, their payment structure, their requirements for length of service and other important details before you sign on the dotted line. If they try to pressure you without disclosing their plan, letting you talk to the person who will be handling your account or sharing their reference, they should be avoided.

+ **Rates based on payment per keyword:** There are many different payment structures for PPC services, but this is one of the ones that doesn't make sense for your budget or the PPC company. The rates for this type of structure are higher if your campaign is more complex (has more keywords). However, your keywords may vary from month to month or week to week. A good PPC manager may try thousands of keywords before finding the right ones, however if you have a strict budget per keyword they'll be limited by how many keywords they can test which can inhibit your success.

+ **No set up fees:** Experienced PPC management companies know that it takes work to get a PPC advertising campaign off of the ground. The time spent on this set up should be billed to the customer. Companies that do not have set up fees are trying to win your business with low rates, but they are normally not experienced enough to provide quality.

+ **They work with your competitors:** This is a conflict of interest and will cause your PPC company to be competing with itself for clicks. Your PPC company should use your competitor's site as a gauge for where your PPC campaign is going but it should not be servicing them as well.

GENERAL PPC QUESTIONS

✓ **Who will be handling my account?** If you're working with a "one man shop", the answer to this question will be simple. However, if you're working with a larger PPC company you may often be speaking with a sales rep before you can communicate with an account rep. They may make you a lot of promises that their PPC experts can't live up to. Ask to speak with a qualified person who will be handling your PPC account before you make a commitment to work with a company. If the sales person refuses to do so, move on to the next company.

✓ **What is the communication process like?** Look for ongoing email support, a 1-800 number and the invitation for open communication. Stay far away from a company that is evasive or puts a limit on how much you can contact your account representative.

✓ **How will my company be involved?** The right answer to this question will depend on your needs. Look for a company that has the right amount of involvement for your experience level and your requirements. For example, if you'd like a company to improve upon an existing group of ads that you have, stay away from a company that wants to start from the bottom up with brand new campaigns.

✓ **What are your qualifications?** A proven track record of results is the only way to truly determine whether a PPC manager will have enough experience to handle your account. Ask for referrals and follow up on those referrals. Make sure that you ask their referrals about the process of working with the PPC management company. Ask the referrals if there was anything they'd change about the experience and if they were happy with results. You should ask the PPC management company about their past successes and especially if there are any clients they've had that didn't get the results they wanted.

✓ **How do you set up your payment structure?** There are a lot of different payment structures out there. We've already discussed the drawbacks of working with a one-time flat rate. These payment structures are more reliable and common among top PPC providers.

The percentage of spend is the most common pricing model. Your PPC company will be paid a percentage of what you spend on advertising. Normally it's around 15%, but the percentages are lower if you end up spending more. With this pricing model it's easy to estimate your charges, but if you spend more the PPC company is rewarded. If you go with this model, you need to have a contract that holds them accountable for the results so they won't apply pressure to spend more in order to get more money.

A flat monthly fee is different than a flat onetime fee. The monthly fee will be based on how much work the PPC company will be putting in on a monthly basis. They will manage and optimize your account on a monthly basis. This is definitely the easiest to budget for but there is no incentive for your PPC company to grow your account. If you go with this option, make sure that your company will be willing to renegotiate in a few months to grow your advertising spread.

Performance based fees are another type of payment structure that is growing in popularity. The reason being is that your PPC manager's success is directly tapped into your campaign's success. This makes them much more responsible for their results. There are normally flat set-up and monthly fees but a percentage of your money is guaranteed back if certain criteria aren't met. The PPC manager will also achieve additional bonuses for having great results. Extra pay is a great motivator and this payment structure can really help you get the most for your money.

✓ **What type of contracts do you offer?** PPC contracts come in one of two varieties. It's important to know which you're dealing with before you commit to working with a company. You'll either be offered a month-to-month contract or a long-term contract. The month-to-month contract is appealing because it is low risk and there is a low up-front investment. However, there isn't much incentive for the PPC manager to get invested in improving your advertising return. They won't use all of their resources on your account because they aren't sure if you'll be with them for the long run.

The other variety is a long term contract, which is normally six months to a year. The PPC manager can become fully invested in your advertising because they can follow through on their testing from month to month. There is a higher investment for long term contracts which leaves you with higher risk involved.

If you end up with a company you are not satisfied with, you'll be stuck with them for a long time. If you are considering a long term contract but nervous about the nature of the service, as if there is an "early out" clause. Most professional PPC companies offer one in their contracts which will allow you to opt out of the contract if you aren't satisfied.

✓ **What are your set up fees?** A lot of work needs to be done with your account before your PPC company places an ad. This front end work is normally charged in the form of set up fees. In addition to asking the amount of the set up fees, they should also make it clear what those set up fees will be covering. There should be specific deliverables for the front end set up portion of their service. For example, they should give you an overview of their ser up process, a keyword list and negative keyword list. They should go above and beyond to spell out to you exactly what they are doing during the set up process.

✓ **How much can I expect for my ROI?** This is an important question but the answer comes with a grain of salt. No company can guarantee an exact return but discussing possible return on investment is important in opening honest dialogue between you and your PPC manager. A professional PPC manager should be able to gather information from your business and give you a snapshot of the risk vs. the reward for your specific business. You'll need to discuss with them your gross revenue, you gross profit, your net profit and the kind of return you can get on your advertising investment. If your website deals with lead generation you need to share how much you can afford to pay per lead, you lead cycles, what lead-to-sale conversion you need to have to make a profit.

You can have the PPC company sign a non-disclosure agreement if need be, but information like this must be shared. Your potential PPC company should develop a risk-reward report for you based on your financials that will help you understand what they can do for you.

✓ **Do you have the resources to effectively manage my company's projects?** If you are a mid to large sized company with complex advertising needs you'll need a company that has the man power and experience to help meet your needs. Ask for specific examples of their experience with companies of your size or in your industry.

✓ **Is your company certified?** Google Adwords certification is a series of tests that ensure that an individual has the capabilities to handle an Adwords account and effectively use pay per click marketing. Among the requirements are managing an Adwords account for at least 90 days, maintaining at least $1000 total spend for the 90 day period and passing a series of official exams. Google certification should be a minimum standard.

✓ **Will you offer a money back guarantee of some kind?** PPC marketing can be very unpredictable so it's hard to find a guarantee. However, very strong PPC managers may offer a performance based guarantee. You'll normally have to commit to a long term program with the PPC company. The upside is that your PPC manager is more accountable. They know that if they fail they'll owe you a portion of your money back. Since they are personally invested in your results, they'll be able to deliver better results.

✓ **What do you know about my field?** PPC managers will be experienced in a wide variety of fields however, it's helpful for your PPC manager to be experienced with the function of your website, it not the specific field. For example, if you're a retail website you'll want a PPC manager who has handled product sale campaigns and other ecommerce issues. Other areas can include lead generation, digital e-products and membership sites. Start by finding PPC companies that help with your business function and then find then find some that have experience in your industry. If their experience is highly different than your business, stay away.

✓ **How will my account representative be rewarded?** This is an essential question if you are working with a large company and will have an account representative. If you are speaking with a sales person rather that the PPC manager, ask to speak to the PPC manager so you can get a straight answer to this question. You need to know how they are being motivated to make your advertising work. Do they receive performance bonuses? Do they have a lot of other clients? What will they get if their advertising campaigns for your company are successful? Since they will be handling a lot of your advertising, you deserve to know how they are motivated to partner will you for success. Your PPC manager should be invested in your success.

✓ **How will PPC help me make money?** If you're getting quotes from PPC companies, you know that this form of advertising can help your business but your PPC manager should be able to spell out to you exactly how he or she envisions this happening. Once they gather some basic information from you about the nature of your business, they should be able to deliver a pre-contract to you that will spell out how their actions will help you make more money.

Whether it's through gathering more leads, converting more sales or increasing general traffic, your PPC manager should be able to tell you how you'll profit from their actions. Most large scale PPC companies will offer pre-contract assessment services for a onetime fee. In this pre-contract assessment they'll look into your market, your PPC competition, your PPC account if you have one and your website.

✓ **Does my company information remain confidential?** This is essential for your PPC campaign. The information you are sharing with your PPC company is highly sensitive. You should require an NDA, and your PPC company may require an NDA from you as well.

✓ **Who owns the PPC campaign?** The keyword research, ad copy and campaign set up should belong to your company and yours alone. This leaves you the option to move on with another company if you part ways with this one.

THE COMPONENTS OF SUCCESSFUL PPC MANAGEMENT

Now that you understand the overview of PPC management and which questions to ask at the start of the process, it's time to look at the multiple components of PPC campaigns. As stated before, keyword analysis, ad creation and landing page development are the basic "broad" areas of PPC. Landing page development will be covered later in this book. The components in this section fit into keyword analysis and ad creation, or help to inform those decisions.

These components are:

o   Determining the initial goals for the PPC campaign. This will help your PPC company decide which actions to take.

o   Choosing keywords and managing keyword lists. This is the crux of successful PPC.

o   Creating effective ads. Ad text will ensure that your website gets visited.

o   Minimizing the cost per acquisition (CPA). This is a major goal because it will increase your return on investment.

o   Automation. Campaign automation can help your PPC company streamline the process.

o   Reporting. Your PPC company should report to you on a consistent basis to keep you abreast of the development of your campaigns.

Throughout each section we'll share the basics of what your PPC company should be doing. Understanding the general steps will help you remain vigilant about the steps that your PPC company is taking. By analyzing their methodology you'll be able to determine whether or not they have your best interests in mind.

DETERMINING INITIAL GOALS

There are several components to the initial analysis. First, the PPC company will look at any existing campaigns that you currently have running. In order to facilitate this analysis, you should provide reports from your Adwords, Yahoo Search Marketing, Bing advertising or other pay per click accounts. Unless you are under contract with them, or they sign a non-disclosure agreement, you shouldn't allow them access to your actual accounts. Share your data for your total advertising costs spent up until this point and share your traffic and conversion results, but only in report form.

The second factor that a PPC company will take into account is your competition and their position in advertising blocks. They will determine which companies are your competition is based on the information that you give them and their own independent analysis. They will do keyword searches for your most common terms. Determining the competition will help them analyze your possible return on investment as well as potential keywords.

The most important thing to remember about setting goals for a PPC campaign is that their goals for your company should be quantifiable and measurable. You can't expect your PPC company to deliver on what they promise if you don't nail them down to specifics. Quantifiable goals will help make your PPC experience a positive one. There are several different goals that you can determine with your PPC company. Here are some examples of the major goals that should be considered at this point in the game.

CLICK-THROUGH RATE TARGET

One of the goals to establish with your PPC company is a click-through rate target for your PPC ads. The click through rate is the number of times that your ad is viewed (called "impressions") versus the number of times your ad is clicked on. Click through rates generally range from 1% to 5%. They can be calculated by a simple formula.

Click through rate % =

Total number of ad clicks/total number of ad impressions X 100

Imagine the difference it will make to get your click through rate up just 1%. Increasing your number of qualified leads will increase your website's success no matter what your goals are. Improving click through rate by selecting keywords and improving ad text should be a major priority.

CONVERSION RATE TARGET

Conversion rate is another important factor in determining your goals for pay per click marketing. Conversion is the number of people who click on the ad and then take the required action. The action may be a subscription, a purchase, a registration or another action on the website.

The conversion rate is determined from either the number of impressions or the number of total ad click-throughs. Most PPC companies calculate the conversion rate based on the actions and the ad click-throughs. Conversion rates can vary greatly, from 5% to 20% for the number of people that click on the ads.

> Conversion Rate % =
>
> Total number of actions/ Total number of ad click throughs X 100

RETURN ON INVESTMENT TARGET

Your return on investment or ROI is an important goal. You don't want to spend lots of money on PPC advertising without carefully considering how much you'll be making back. The ROI of your campaigns is determined by the bottom line revenue increase that your company sees from advertising with a PPC company. A company may also see an expense decrease while working with a PPC company. If they have been handling their own PPC accounts and not seeing much in return, a professional PPC company can spend less and get better results.

Your company's ROI is determined by your average profit per sale and your estimated number of conversions. The ROI formula is calculated in two parts. First, you have to determine your contribution.

> *Contribution=*
>
> *(Your average profit per sale x Your Estimated number of conversions) – Advertising costs*

The contribution figure is used to determine the ROI.

> *ROI = Contribution/ PPC Cost*

For example, if your average profit per sale is $150, your number of conversions per month is 75 and your costs are $1500 for the month, here's what the formulas would look like:

> *($150 x 75) - $1500 = $9750 for your contribution*
>
> *$9750/$1500 = 6.5% ROI*

Improving the return on the investment can be accomplished by increasing the number of conversions or adjusting the advertising costs. Setting an ROI goal can help your business become more profitable.

COST PER ACQUISITION

The cost per acquisition rate of your advertising is the amount of money it takes you to acquire a new customer. If your CPA goes down, your ROI will go up. Here's how to calculate the current CPA rate of your website, which your PPC company will use to set a benchmark for your PPC campaigns.

> *CPA Cost=*
>
> *# of new customers each month / cost spent on PPC advertising*

Minimizing CPA is another important part of PPC management. If your number of customers goes up even though you've spent the same on advertising, you instantly have more return on your investment. There will be a section later in this chapter devoted to CPA minimization and how to evaluate whether or not your PPC company is doing a good job.

All of these formulas can be used by your PPC company to determine how your PPC campaigns should proceed. These calculations are important to understand as starting points for your campaigns. You can't set goals and reach goals if you don't have a starting point.

RED FLAGS FOR SETTING INITIAL GOALS

- **Not setting goals at all:** Promises to improve your business with PPC without having measurable goals is a major red flag for a PPC company. Experienced companies know that in order to achieve results they have to set goals and be able to measure their progress.

- **No initial evaluation:** Your PPC company should offer an initial consultation where they take your statistics and use the previously outlined formulas to determine the current state of your advertising.

- **Not asking about your call to action:** A professional PPC company should start with your call to action and then work backwards from that statement. They need to understand the purpose of your site and let you know how that call to action specifically relates to PPC management goals.

QUESTIONS FOR SETTING INITIAL GOALS

- ✓ **What constitutes success with PPC?** You need to know what the PPC companies considers to be a success, because it might be much different from your goals. Look for a detailed answer based on your previous data, and not a general statement about increasing your business.

✓ **What techniques will you use to achieve my PPC goals?** This should go beyond a general statement of starting campaigns and looking up keywords. This will give you a peek at their methodology so you can determine if they are on the right track.

A qualified PPC company should be able to answer these questions and present you with a comprehensive plan to increase your traffic and increase your sales.

Case Study #1: Stein Diamonds

SEO Company: WPromote (http://www.wpromote.com)

Stein Diamonds (www.steindiamonds.com) is a wholesale jewelry manufacturing leader. They sell certified diamonds, luxury watches, earrings and pendants. Stein Diamonds was able to break through barriers and get customers to purchase jewelry online, which is unique for the industry. Most jewelry customers like to browse online and then purchase products in person. Despite breaking through the online purchasing barrier, Stein Diamonds wanted to reduce their Google Adwords campaign spending and maintain their existing conversion rate.

Case Study #2: Light Bulbs Etc!

SEO Company: JumpFly (http://www.jumpfly.com/)

Light Bulbs Etc! (http://www.lightbulbsdirect.com) had been running their own Google Adwords and Yahoo! Search Marketing campaigns for two years previous to hiring JumpFly. Although their PPC campaigns had seen some success, they were not satisfied with their results. They were experiencing high cost conversions and needed to expand their PPC reach, and wanted an experienced PPC firm to take over this form of advertising.

KEYWORD SELECTION

Once the goals for your PPC advertising are established, it's time to begin with keyword selection. The keywords are your bidding tools and they form the crux of your PPC campaign. In the hands of a capable PPC company, your campaign will be based around keywords that are profitable and that bring you new customers. Your PPC company can also help uncover keywords that will be helpful in determining SEO optimization keywords.

Your PPC company will start with the most common keywords for your industry and then begin developing a more customized list based on your particular advertising needs. Keyword selection is crucial in identifying the right market and putting your ad in front of the right group of buyers. Ineffective keyword selection decreases your ROI and costs you more in advertising.

Keywords are phrases or words that people use to look up information online. When you use a search engine to locate information you are entering keywords to help navigate through the vast amount of information online. All search engines record keyword usage, which can be used by advertising professionals and business owners to determine what people are looking for.

Your PPC company will determine which words and phrases are essential for your target audience and what will be most effective to use to advertise your website. They will use several methods to determine the appropriate keywords for your website based on the needs of your website and the current competition. Whereas with SEO, a provider will seek popular phrases for your website and choose the most important terms for optimization, a PPC provider also has to look at the competition. If there is too much competition for a particular keyword phrase, the cost to get the ad in the first three or four ad spaces will be too great to justify that ad placement.

You want your ad to appear in at least the top three ad positions in order to be the most effective. PPC is all about visibility. Average users do not make it past the first pages of results. Most do not look past the first half of the page when trying to locate information. Being toward the top of the search engine results is the clearest shot you have at getting clicks that convert.

It's a delicate balance between finding keywords that will convert and keywords that will fit into your ad spend budget. The formulas in the previous section will help inform your PPC company's choices. If they know your visitor value number is $2.00 they won't bid more than $2.00 for any particular keyword.

The following steps should be used by your PPC company in order to develop a keyword list for your PPC campaigns. This list should be between 100 and 500 quality keywords to start with, and additional words should be added as your campaigns begin returning results. Keyword discovery is an ongoing process that needs to be refined continuously for best results.

DEVELOPING A KEYWORD LIST

Your PPC company will begin to develop a long list of keywords related to your market. That's why it's essential to find a PPC company that has familiarity with your industry. An experienced PPC manager can pull in keywords and phrases that others may not be able to identify. In addition to general brainstorming, your PPC company may also review your log files to see what terms are leading customers to your website.

Your PPC company will also look at your competition to see what terms they are optimized for and what terms they are receiving traffic for. Reviewing the direct competition can include taking a look at their META data and using an external tool like Spyfu (http://www.spyfu.com). These tools will help expand your PPC company's list for your site. They will also use general keyword research tools like Wordtracker.com or Google's Adword Search tool to further expand the list.

FOCUSING THE KEYWORD LIST

There are many keywords that can be used to describe your site, but only a specific type of keyword will get you the results you're looking for. In addition to identifying the relevancy of keywords and the popularity of the keywords, your PPC company will also focus on conversion for your keyword lists.

There are different types of keywords that people use to find information online. They use these types without realizing that they are using them, but their usage can be a good indicator of what they are planning on doing with the information. By using the right kind of keyword phrases you can find visitors who will convert to buyers.

Your PPC company should focus its efforts on finding keyword phrases that relate to making purchases. A person looking for "Office Star Screen

Back Mesh Seat" is closer to making a purchase than someone who is just looking for "office chair." The more specific the keywords are, while still remaining popular, the better.

In addition to looking for specific model numbers or product names, your PPC company should also add purchase specific keywords, like "buy black office chair" or "best price on recycled copy paper". These keywords indicate that the searcher is ready to make a purchase. Although they aren't as powerful as product specific keyword bidding, they are still more likely to bring in actual sales than the broad phrases used by people who are just looking for information.

More detailed keyword phrases lead to better results, which can help your PPC company meet your goals more quickly and for a lower price. By using more specific terms:

o   The click through rate will increase because the search terms and ad copy are closely related to a product or service available on your site.

o   The higher click through rates are actually rewarded by Google. Google rates ads on their relevancy. If an ad gets a lot of clicks it is deemed more relevant by Google. The more relevant an ad is, the more likely it will be placed higher in the ad block.

o   Your conversion rate will increase.

o   Your cost per conversion will decrease. Long tail terms are normally cheaper per click than broad, general terms.

A final step in determining the keyword list for your website is your budget. PPC managers will evaluate how much each keyword phrase will cost for bids. They will relate that to traffic each keyword will potentially receive and the relevancy of those keywords to your website. They should determine where your money is best spent for your keywords.

KEYWORD MATCHING

After determining what keywords you should be bidding on, your PPC company should decide on the targeting match for your campaigns. How should those keywords be related to user inquiries? For example, should your ad for "Xerox Multipurpose Recycled paper" show up for "Xerox multipurpose paper" and "Xerox recycled paper" as well as that exact search term?

Your PPC manager should determine if the ads should show up for exact match, phrase match or broad match queries to the search engine. Their decision will make a big difference in how effective your ads are.

o   *Exact match*

This category is referred to as "Standard Match Type" in Yahoo search marketing and "Exact Match" in Google Adwords. When this type is selected, your ad will only be displayed when the user searches for the exact phrase that you are bidding on. It is the most highly targeted of all the options available, and is a good place for your PPC company to start your campaigns. Once they've established basic metrics with exact match, they will open up your bidding to broaden to other match types. Exact matches are denoted with brackets, as [black office chairs].

o   *Phrase match*

This type of match is used when a user queries a related term that includes all of the words in your keyword phrase. For example, Google will display your ad for "Xerox multipurpose recycled paper" for search queries like "buy Xerox multipurpose recycled paper" or "Xerox multipurpose recycled paper ream." Phrase matches are displayed with quotes, as "black office chairs."

o   *Broad match*

Broad match is the largest match group, but it can also be the most costly. With this setting, your ads will appear for any search terms that have ANY of the words in your selected keyword. Your ad for Xerox multipurpose recycled paper will show up for Xerox paper, Xerox multipurpose, multipurpose paper, recycled paper and even simply paper.

The benefit of going with this keyword matching group is that it is includes plurals and other combinations your PPC company may not have thought of. However, it does increase the costs of your PPC campaign because your ad appears in front of less targeted searchers. It shouldn't be used initially but can be a helpful tool later on in order to gather more search terms. Broad match keywords have no special punctuation.

These options should be utilized by your PPC manager in order to get the best combination of displays and results. Be sure to login to your own PPC account to determine if your PPC manager is using the right type of selection for your keyword phrases. If all of your campaigns are running on broad match, you can end up spending a lot more on PPC than you need to.

ELIMINATING NEGATIVE KEYWORDS

Another tactic that your PPC company should be using when refining your keyword list is eliminating negative keywords. This is especially important if they are using broad keyword matching or phrase matching. The negative keyword filter that is in most PPC search engines will allow your PPC company to further specify which keywords your ad should be displayed for.

For example, your PPC company may decide to run a broad match campaign for "recycled paper" for your office supply store. However, you sell recycled printer paper and multipurpose paper, and not recycled stationary or recycled paper bags. These two phrases are related and they may display your ad if the negative keyword option is not used.

Your PPC company should eliminate any negative keywords that may display your ad that aren't directly related to what you have to offer. They would choose recycled stationary paper and recycled paper bags as negative keywords, in addition to any other terms that have the words "recycled paper"

Negative keywords should be used if your PPC company is utilizing the broad match category for your PPC campaigns. It allows them to have a broader reach with the list and discover more keywords without you having to pay for those negative keywords.

Your PPC company can use negative keywords to fine tune your bidding strategy. In addition to this tactic, they should be moving keywords through different match types in order to test their effectiveness. For example, if an ad is doing particularly well as a phrase match ad, they should create a broad match ad to see if there are any other profitable phrases that they can "trap" with this broader keyword phrase group.

Match types are a profitable but often overlooked aspect of PPC marketing. A good PPC company will be able to use the exact – phrase – broad strategy in order to optimize your keyword list and increase your ROI.

RED FLAGS FOR KEYWORD DISCOVERY

+ **Using a long list of untargeted keywords:** Your PPC company should be focused on words that will convert and not just use long lists of related keywords. Generic keywords that are selected based on traffic and not on your individual business needs will cost you a lot of money in the long run. Ask for a list of keywords before your PPC company starts the bidding process so you can be sure that they are targeting the right keywords.

+ **Using keywords that are unrelated:** On the other end of the spectrum from untargeted broad keywords are keywords that are specific but they don't directly relate to your business. The term "recycled paper stationary envelopes" may get a lot of traffic and bring people to your office supply website, but if you don't offer that product you'll end up paying for those clicks without getting any sales in return.

+ **Using too few keywords:** A short keyword list from your PPC company is a sign that they aren't doing enough research on your part. Ideally, your PPC company will start with a long list of keywords and then refine that list based on the results. They will eliminate keywords that aren't effective and then focus on making the ad groups that are working convert even better.

+ **Not updating your keyword list with emerging keywords:** Emerging keywords are phrases and terms that may grow in popularity after your PPC company has developed your initial keyword list. PPC campaigns become more refined with time and if your PPC company is not looking for emerging keywords and adding them to your campaigns you may be missing out on valuable traffic.

+ **Not determining the unique aspects of your business:** A large part of your targeted keyword list should be related to the uniqueness of your business. If you have keywords and campaigns directly related to your USP they'll be more likely to convert well. By looking at your competition, your PPC company should be able to identify what makes your company different and use keywords that zero in on those differences to make you stand out.

+ **Bidding broadly without negative keywords:** When your PPC company begins targeting keywords, they should focus on exact matches or phrase matches. Setting up all of the campaigns on broad matches alone is lazy and costly. Without using negative keyword phrases you'll end up paying a lot more for your PPC campaign without seeing results. Broad match keywords should have an extensive list of negative keywords, or not be used at all.

+ **Avoiding geo-targeting:** PPC search engines allow your PPC company to target ads to local areas. You can get as specific as a state for your PPC ad displays. Even if your company is selling all over the world, your PPC company should come up with ads that target specific states. An ad with a state name detail will be better received than the same ad with a general term. Using geo-targeting shows the prospects that you recognize their location and it will increase your click through rate.

Avoiding bids on your company name: Your company name is a keyword, and it's the most targeted keyword that exists for your website. Your competition may try to target your company name as a keyword, so you should as well. Even if your company is not well known for its name, it doesn't hurt to have your business name used as a keyword. You might as well pay for the advertising especially if there isn't a lot of competition for the keyword phrase.

CASE STUDY #1: STEIN DIAMONDS

Stein Diamonds already had PPC marketing campaigns in place when WPromote began working with them. Their attention was focused on pulling back advertising on keywords that weren't converting well and adding new product based keywords.

For example, our case study Stein Diamonds has the first ad spot for its name, but other diamond retailers are also using the keyword "stein diamonds" for bidding.

QUESTIONS TO ASK ABOUT KEYWORD RESEARCH

✓ **Describe your keyword research process:** Your PPC company should explain that they select at least 100 to 500 highly targeted phrases to develop your PPC campaign. They should also note that they will be watching for emerging keywords that may come on the radar later on during their service period. They should describe using niche statistics, competition and relevance to find keywords that will convert for your website.

✓ **Will you use the keywords that I am already using?** Your PPC company should utilize keywords that you are already using for your PPC campaigns and other advertising, but they should also let you know if those keywords aren't effective. If you haven't done sufficient keyword research for your advertising your PPC company may let you know that your choices aren't profitable. Be open to accepting completely new keywords for your terms, but your PPC company should use your list as a starting point.

✓ **What types of keywords will you use?** Look for an answer that reflects the different ways in which people use Internet search queries – for information and for purchase. They should respond that they focus on purchase related keyword terms that will increase the likelihood of resulting on purchasing traffic.

✓ **What keyword match types will you use?** Exact and phrase matching should be used first rather than broad match types. These can improve accuracy and save you money on advertising. Broad match types should only be used with liberal use of negative keywords.

✓ **What percentage of keywords will be based on our brand name?** Your company name should be an essential part of your PPC's advertising approach. However, the amount of money spent on your brand name will be determined by how many people are specifically looking for your brand name. If your brand name is not a popular search term, there should be one ad group and perhaps just one keyword used. However, if it's a more well known brand your PPC company may suggest bidding on keywords like "your company name reviews" or "your company name

✓ **How will you implement negative keywords?** Look for answer that shows a complete understanding of negative keywords. Ask them to give you a few examples of negative keywords that they'd use on popular keyword terms for your niche. They should show you that they understand which terms would be most profitable for your site.

Case Study #2: Light Bulbs Etc!

Light Bulbs Etc! was using PPC marketing on their own, so there were conversion stats for specific keywords. JumpFly used the previous stats to develop a keyword list and then added additional keywords based on their budget. They added Microsoft adCenter pay per click marketing campaigns for additional keyword coverage.

Case Study #3: All Things Aquarium

All Things Aquarium was brand new to online advertising. The account manager at JumpFly spoke with the business owner to obtain information about their business. A keyword list was developed based on their geographical area and aquarium related keywords.

CREATING EFFECTIVE ADS

Once the keywords are in place, your PPC company will begin to develop your ads based on your needs and what your company has to offer. Your PPC company only has a few lines to grab the attention of the website visitor and get them to visit your website. Without the right copywriting skills and ad targeting, your impressions will stack up and your quality score will drop.

Some business owners may think that showing untargeted ads that don't get clicks is no big deal. After all, if your ad is shown but not clicked on you don't have to pay. However, Google and other pay per click search engines actually penalize you for showing unrelated ads. If your ad for "Xerox copy paper" shows up for "Xerox copy machines" and is not clicked on, this effects your quality score.

If your quality score is lowered, your ads won't show up as often. Your account may be labeled as irrelevant and your ads across the board may not receive as high ad placement. That's why it's so important to make sure that the ads are showing up for the right keywords, and that the ads themselves are highly relevant to the keywords.

By using ad groups you can target your marketing toward specific groups of people. Ad groups for an office supply store PPC account may include an ad group for paper, an ad group for office chairs, an ad group for accounting software, etc. Your account with have several campaigns, each with its own ad group which in turn has its own set of keywords.

Here is the basic set up of a PPC account:

Look for this structure when you log into your PPC account after your PPC manager has been handling it for a while. There should be several campaigns that target broad keywords, but within those campaigns are hyper targeted ad groups that relate to specific products or long tail keywords. The keywords that are group under each ad are closely related. For example, an ad campaign for office chairs might have three ad groups – "ergonomic office chairs", "manager office chairs" and "black office chairs."(Of course a real life campaign for office chairs would have dozens more ad groups depending on the keyword research).

Within each of these ad groups, your keywords will be variations of the main keyword. The ad group "black office chairs" would include the following keywords.

"black office chairs"
[black office chairs]
"black office chair"
[black office chair]
"black mesh office chairs"
[black mesh office chairs]

It would also include any other closely related terms. The reason they need to be closely related is because your PPC company will be using the same ad for each ad group. The ad in an ad group needs to make sense for all of the keywords in that ad group. "Leather chairs" or "ergonomic office chairs" wouldn't be in this group because displaying an ad that referred to "black office chairs" wouldn't make sense for those keywords.

Separating each category of product into a different campaign will also help your PPC company manage your account better. With just a glance they'll be able to see if specific ad groups are doing better than others. For example, if the ads in the office chair campaign aren't doing as well as the ads in the accounting software campaign, they may move more of your financial resources to the software campaign. In addition, if there are certain ads within that software campaign that are outperforming others, they will expand the ads to those keywords to take advantage of the high click through or conversion rate.

PPC AD CONSTRUCTION

If you want to evaluate the quality of the ads that your PPC company is using for your campaigns, you need to know about the basics of PPC ad construction. The main goal of a PPC ad is to get the viewer to click on the ad. It has to take them out of searching in the free results and get them to click over to your website.

Advertising of all types needs to grab the reader's attention, and pay per click ads are no different. Research shows that most people reading search engine results page spend most of their time focused on the left hand side of the page, away from the paid advertising. In order to grab the attention of the target market your PPC company should focus on two things - incorporating keywords and relating the ad to the market's needs.

There are four lines to every PPC ad : the headline, two lines of copy and the link to your website. Typical ads look something like this:

From Google: From Yahoo:

Modern **Office Chairs**
We have the best selection of
modern **office chairs** on the net!
www.CoolChairz.com

Black Office Chairs
Shop With Confidence! We Offer
Lifetime Warranties on All Items.
OfficeFurniture2go.com

As you can see, both pay per click search engines have the same structure – headline, two lines and the URL. The top headline is a clickable link that is connected to your website. The bottom URL is called the display URL and can be set to any URL you like. So for example, the top link could be connected to http://www.yoursite.com/blackofficechairs.html and the display URL would be set to YourSite.com.

Your PPC company will be in charge of developing your ads, but you should look for the use of keywords in the headline and a message that speaks directly to the market. If a person is looking for a black office chair, they are more likely to click on an ad that says "black office chairs" in the headline. In addition, Google and Yahoo both make keywords bold if they are used in the headline.

In the examples on the previous page, the keyword phrase was "black office chairs" and you can see that "office chairs" has been put in bold in the Google ad and that "black office chairs" is in bold in the Yahoo ad. This is an attempt to attract the reader's attention. If they notice the phrase they are looking for is in bold, they'll be more likely to click.

That's why when your PPC company organizes your ad groups, they should keep related keywords in the same ad groups and give other variations their own ad group. Placing all "black office chairs" related keywords in the same ad group will mean that the ad for that group will closely relate to those keywords.

buy black office chair
office black chair
black mesh office chair
black office chairs
black office chair

Black Office Chairs
Choose Your Perfect Chair Today
Free shipping, low prices
www.yoururl.com

Keyword usage is one way to attract the attention of the searcher, but speaking directly to them is also very effective. No one likes to be sold to, but if your PPC company can show the searcher what is in it for them, they are more likely to click on your ad. Searchers want to know what they'll get out of clicking on your ad. It might be a lower price, a quality product or just a feeling. Whatever the motivation is, your PPC company needs to nail it down and convey it through the small ad space. It all starts with listing the benefits of what the searcher will find on the other side of that click.

The landing page will determine what benefits your PPC company will convey to the web searcher. People are interested in what is in it for them. Your PPC company can use the phrases "you" or "yours" in the ad to attract attention and emphasize their relation to the reader. Take a look at the difference between these two ads-

Black Office Chairs	Black Office Chairs
Lots of choices, low prices	Choose Your Perfect Chair Today
Many to choose from	Free shipping, low prices
www.yoururl.com	www.yoururl.com

The second ad uses "your" to connect with the audience. It also invokes urgency, another important tactic in copywriting. By using the phrase "today" your audience will be more likely to take action now.

Your involvement in the creation of the actual PPC ads will probably be limited to directing the PPC company to your landing pages and helping them develop a benefits list. However, with access to your PPC account you will be able to take a look at the type of ads the company is using. Make sure they look more like the second example than the first

RED FLAGS FOR CREATING EFFECTIVE PPC ADS

+ **Ineffective headlines:** The headline of a PPC ad should capture attention and give the reader a reason to click on the ad. There are only so many characters available for your PPC headline. Your PPC copywriting team should use these characters wisely.

For example, the headline should not state the name of your business. The URL states the name of your business, so it's

a waste of space to repeat the business name in the headline. The reader doesn't particularly care about your business name – they want to know what is in it for them!

Fuzz Productions
Exceptional **Web Design** and Custom
Interactive Marketing Services.
FuzzProductions.com
Massachusetts

(source: http://www.wordstream.com/blog/ws/2009/10/13/ppc-ad-writing-lessons)

Headlines should include the keyword that is being bid on. This increase the relevance of the ad and can increase the position of your ad. Experienced PPC copywriters will be able to engage the reader and use the keyword at the same time. For example, this ad is displayed for the keyword term "heart monitor":

EKG **Heart Monitor** Service
No Cost EKG **Heart Monitor** Services
Holter & Event **Monitors** Provided
www.cardiolabs.com/Heart_Monitor

(source: http://www.wordstream.com/blog/ws/2009/10/13/ppc-ad-writing-lessons)

The ad uses the keyword in the headline and also provides additional details in the space. Look for a similar use of keywords in the headlines created by your PPC company.

+ **Repetitive keyword use:** When an ad is used in the PPC systems under a specific keyword, that keyword is made bold. Some PPC advertisers mistakenly take this as a cue to use the keyword as much as possible. Including a keyword will help your PPC ad's relevance, but it is not necessary to repeat the keywords several times. It's a waste of advertising space.

(source: http://www.wordstream.com/blog/ws/2009/10/13/ppc-ad-writing-lessons)

- **Improper campaign structure:** Ads get more exposure if they are relevant to the keywords your PPC company is bidding on. Once your company has set up your campaigns and is running ads, login into your account and check on the campaign structure. Be sure that the structure mimics the example outlined previously, with each main keyword separated into a campaign group. A poor PPC structure will hurt your campaign. For a qualified PPC company, this will be a natural part of their operation. Remember, multiple campaigns will get you better results.

- **Not using multiple ad versions:** PPC services allow your PPC company to create several ads for the same keywords. The PPC service will rotate the ads so that they are all displayed. This will give your PPC company valuable data on the effectiveness of their ads. They will use this data to create more refined ads to get you a better ROI for your PPC advertising. Ultimately, if a PPC company is not using the split testing capabilities of the PPC services, or an external tool for split testing, they are wasting your advertising dollars.

  Here is an example of a split test for an ad created for wireless headphones. Please note that the headlines are written in an automated code that will fill in the keyword variation for that particular campaign. When the ads are run, the headline will appear as "Samsung Wireless Headphones" or "Casio Wireless Headphones". The copywriter is testing the best feature of the wireless headphones to see what readers are drawn to the most.

{Keyword:Wireless Headphones}
Clear Reception From $29.99
Free Courier Shipping
WirelessHeadphones.com

{Keyword:Wireless Headphones}
Great Selection From $29.99
Free Courier Shipping
WirelessHeadphones.com

{Keyword:Wireless Headphones}
Stylish, Wicked Bass From $29.99
Free Courier Shipping
WirelessHeadphones.com

{Keyword:Wireless Headphones}
Name Brands From $29.99
Free Courier Shipping
WirelessHeadphones.com

Case Study #1: Stein Diamonds

WPromote incorporated new ad text into existing ads the company had running. They analyzed day trends and utilized split testing to make their ads even more effective.

Stein Diamonds ad appearing for "Chopard"

Chopard Watches
Wholesale Prices Lifetime Warranty
30-Day Guarantee Free Shipping
SteinDiamonds.com

(source: http://www.getelastic.com/ppc-split-test-strategies/)

+ **Not pointing out unique aspects of your product or service:** In the previous example, the copywriter was testing various aspects of the headphones to see which appealed to the reader the most (the clear reception, selection, bass output and name

+ brands). Your PPC company should include unique features of your business and your products or service in your PPC ads. If the ads are generic, they will get lost in the sea of other ads out there and your conversion will plummet.

+ **Not linking to the landing page:** Remember that PPC ads offer a display URL and a link URL, which can have different text. If the landing page for black office chairs is http://www.bestofficeproducts.com/black-office-chairs.html your display URL can be www.bestofficeproducts.com and the traffic will still route to the former link. Your ads will convert better if the traffic is routed to the landing page that is specific for the bid keyword.

+ **Not optimizing the landing page:** The landing page is essential to making PPC conversions work. Your PPC company should be able to give you specific advice on how to optimize your landing pages so that your PPC ads are working the way that they should be.

+ **Using telephone numbers in the ad:** Telephone numbers may appear to be eye-catching, but they are a waste of valuable ad space. If people are searching online for your business, they are not interested in picking up the phone and calling your business. If a PPC company uses your phone number in an ad, they are inexperienced.

QUESTIONS TO ASK ABOUT CREATING EFFECTIVE PPC ADS

✓ **How many years of experience do you (or your copywriters) have with PPC ad creation?** Knowing how to create effective PPC ads takes time. Although testing will refine the process and get you better results, it helps to have the first draft of the ad be effective from the start. Years of PPC writing experience will give your PPC provider the skills that they need to start getting you results from the start. This will save you money and give you a higher ROI.

✓ **What is your approach to ad copywriting?** They should respond with a sound strategy based on experience. Their formula should include all of the points outlined above, especially with regards to finding the benefits and features of your product or service. An experienced copywriter will be very familiar with promoting the benefits. The approach should also include split testing and ad refinement.

✓ **How will you organize my PPC accounts in order to create effective ads?** This question will test their knowledge of how to structure your account for the best ad coverage. They should respond by saying that each keyword group will be given its own campaign so they can custom tailor the ads for that group to the keywords.

✓ **How many versions of an ad will you use for each keyword?** The correct answer is that they will use as many as it takes in order to maximize the effectiveness of the ads. They should explain that they use split testing features to make the ads more effective on each round of publication.

Case Study #2: Light Bulbs Etc!

JumpFly incorporated ad copy testing based on previous conversion tracking data. They also tailored the ads to more specific landing pages based on each keyword term.

Light Bulbs Etc! ad appearing for "light bulbs"

Light Bulb Superstore
Low Wholesale **Light Bulb** Prices!
Complete Line of Lighting Supplies.
www.LightBulbsDirect.com

Case Study #3: All Things Aquarium

JumpFly used geographically targeted keywords to create ad campaign groups. Since the company was new to PPC, they created brand new ads to be utilized and tested.

All Things Aquarium ad appearing for "salt water aquarium Chicago"

Saltwater **Aquarium** Pros
Custom Design, Setup & Maintenance
Serving Chicagoland for Over 21Yrs.
www.AllThingsAquarium.com
Chicago, IL

CPA MINIMIZATION

Once your PPC company has researched keywords, developed campaigns and created ads for your business, their next goal is to lower the amount you are spending per click. This is where an experienced PPC company can really prove their worth. You may be able to get traffic from your own PPC efforts but your company will probably spend a lot of money to get those clicks. A PPC firm can take your ad spending and increase your ROI so you are spending less to get more conversions.

The cost per acquisition, as outlined above, is the number of new customers obtained by your company divided by the amount your company is spending on PPC each month. There are several different methods that a PPC firm can use to minimize your CPA. Some of them we've already touched upon in the keyword selection and creating effective ads sections. Bidding on specific keywords, creating ad channels and routing traffic to landing pages (rather than the main pages) will help minimize cost per acquisition.

PPC AD POSITION AND BIDDING STRATEGY

The first major step you should look for from your PPC company is your PPC ad position. The first ad position in the search engine results should not be your goal. The best volume and ROI combination can come between the second and fifth position. If you have been concentrating your bidding on getting position #1, your PPC company can instantly minimize your CPA by dropping down your position.

Dropping your company's ad position is an example of a bidding strategy that your PPC company can use to increase the ROI for your pay per click spending. There are two different approaches to bidding strategy that your PPC company can use. The first is a steady strategy focused on slow growth. It's appropriate if your company has a limited advertising budget and lots of time for testing and refining.

The steady strategy involves selecting thousands of keywords that are appropriate for the niche and the website. Your PPC company will set the bids just high enough so that they are getting a decent position, normally between position 4 and 8. This way the top performing keywords are within the top page without the high costs associated with maintaining an ad placement of 1 through 3.

After placing the ads, your PPC company will look over the multitude of keywords and see which ones are producing conversions. They will use your traffic statistics and sales figures in order to determine which keywords are converting. The bid price for these keywords is increased until the bid price approaches your maximum cost per click.

The major advantage of this steady strategy is that your PPC company will limit your risk of overspending on keywords that aren't converting. However, it may take weeks or months to determine which keywords are converting the best. This can be frustrating if you want more instant results.

The second strategy is called the aggressive strategy. As the name implies, this strategy is bold and it requires a larger PPC advertising budget initially. Your PPC company will bid high on all of your ad groups and your keywords. They will aim to get your ads within the first to third positions for all keywords. This will give them a lot of data on click through rates and conversions very quickly. They will take this data to incorporate additional keywords in successful ad groups.

The aggressive strategy costs more but there are several benefits. The first is a rapid return on results. With aggressive bidding, your ads will be in a position where they can collect data quickly. In some small niche markets, there may not be very many people searching for certain keyword combinations. With the aggressive strategy your ads will appear first and it will increase the likelihood that your ads are clicked.

In addition, having an aggressive strategy will increase your quality score. When your ads are placed higher they are more likely to be clicked. The more clicks you get, the more relevant your ad appears to be with regards that that niche. It will increase your quality score meaning that all future ads will be placed higher automatically. This is often referred to as the "halo effect" and it will persists even after your PPC company begins lowering your bids.

If your PPC company utilizes an aggressive strategy, you can expect your CPA costs to be higher initially. Once the PPC team determines where your money is best spent, they will begin to minimize the CPA and increase your results. It's the same process that a steady PPC strategist will use, but the decisions are made much more quickly.

IMPROVING QUALITY SCORE

In the previous sections, we've referred to the Quality Score system that Google uses to help determine ad position (in conjunction with the bid price). The Quality Score system is exclusive to Google, but PPC experts have noted that the techniques used to improve quality score will also help with other pay per click search engines.

Quality score is determined by a number of different factors, some of which are kept hidden by Google. Even though there are factors to the quality score that are kept secret, your PPC company will be able to take a lot of steps to increase your quality score. This is one of the most important factors to minimizing CPA costs. An account with a high quality score will consistently get better placement for their ads than an account that hasn't paid attention to quality score factors.

For example, two companies have ads for "black office chairs." The first company bids $0.50 per click and has a high quality score. The second company is brand new to PPC and bids $0.75 per click in attempts to get to the top of the results. However, because of the quality score factor, the first company may get higher placement (and more clicks) even though their bid is lower. If your PPC company works to improve your quality score, you'll be able to get better results from lower bids.

Google breaks down quality score on a per ad basis. Within the management dashboard of your Google Adwords account, your PPC company will be able to check the quality scores of the keywords that are being used. They will normally pause keywords that have quality scores of four or lower in order to improve the quality score of your account as a whole.

SEARCH VS. CONTENT NETWORKS

This is a Google specific area that needs to be addressed by your PPC company. Google offers opportunities to advertise in its search engine results, as shown in earlier screenshots. However, your ads may also be shown in the content networks of Google. Google offers website publishers the opportunity to display ads on their sites with the Google Adsense program. Ads are displayed based on the nature of the content on the page. Here's an example of the content network advertising program:

(source: http://bestcustomgolfballs.com/)

When your PPC company sets up your Google Adwords campaigns, they have three choices. They can choose to have your ads displayed in the Content Network, in the Search Network or in both. From a layman's perspective, getting the most exposure seems like the best tactic. However, there are benefits and drawbacks to advertising with each network.

The Content Network will display your ad if the page in question has similar keywords. The matches don't have to be exact. In the previous screen shot, the page of content discusses golf handicaps and the ads that are displayed are about golf jobs, improving golf club swings, Nike golf balls and custom fitted clubs.

As you can see, the results are much less targeted than they would be with search results. If the ads were being displayed on a search page for the term "golf handicap" the ads would all include the keyword "golf handicap" and would be closely related to the term.

Sponsored Links

USGA Golf Handicap
Official USGA **Handicap** $29.95
Online Stat Tracking. Local Events.
Golfsmith.USGCN.com

Establish a **Golf Handicap**
USGA, CONGU, AGU compliant **handicap**
Sign Up for Free! anyone can join
www.GolfFrontier.com/Handicap

USGA **Golf Handicap** Online
Trending Reports, Analysis Tools,
Graphs, Easy to use - Free Trial
www.NetHandicap.com

Search engine results pages account for only about 5% of websites (Geddes), so advertising in the Content Network will expose your ads to a larger group of searchers. The more your ad gets in front of targeted web searchers, the more likely you'll be able to get clicks and get conversions. However, content behaves a lot differently than search does. When someone is browsing through a website they've already taken a step closer to finding the information that they are looking for. They've arrived on the page through search and may not need the product or service you are displaying in your ad.

In addition, your ad may not directly relate to the information that the searcher is looking for. If they are searching for the lyrics to "Sugar Pie, Honey Bunch" and find a page with the lyrics, your accompanying ad for raw honey isn't going to have any relevance to their needs. An experienced PPC company can limit the tendency for your ad to show up on completely unrelated Content Network pages, but the rules are much less exact than they are for the Search Network.

Many experienced PPC companies will begin with a Search Network campaign and then extend it with a completely separate campaign that uses the Content Network. This helps minimize CPA costs because the Search Network and the Content Network display ads differently. Since limited keyword matching occurs within the Content Network, you could end up paying a lot more with advertising if the bids are the same for both networks.

Your PPC company should either eliminate the Content Network initially or structure your account in such a way that there are separate campaigns for each network. Within the search campaign, you can set a high budget and your PPC company will determine bids by your return on investment per keyword. The content campaign will have a low to medium budget. This will help your PPC company determine which sites where your ad is being displayed are converting well. Finally a placement campaign can be created that will show your ad only on the content sites that have been proven in the past. (Geddes)

GEOGRAPHIC TARGETING

PPC ads can be categorized by the geographic area in which they will be shown. As a result, your PPC company can minimize your CPA by only showing the ads in relevant geographic locations. Not only can they research keyword terms that are related to the geographic area in which a company is doing business, but they can arrange is so that the ads are only displayed for particular geographic areas. If you have a locally based business that needs traffic from the community, and not clicks from halfway around the world, your PPC company can target your campaigns geographically and increase your ROI.

RED FLAGS FOR CPA MINIMIZATION

+ **Inefficient ad spend:** PPC is one of the areas where hiring an inexperienced provider can really cost you a lot of money. You should understand that it takes time to develop the correct keyword list and optimize your PPC account for conversions. However, after a month to six weeks you should be able to see some positive results from your PPC efforts. Widespread spending across the board with no increase in conversions a sign that the PPC company is wasting your money, not to mention harming your future Quality Score Index.

+ **Not changing ads or keyword groups:** An effective PPC campaign (or set of campaigns) is always moving. The ads are always being refined and the keyword lists are always being tweaked so that you get the highest number of conversions for the lowest price possible. An inexperienced PPC company will let things run to long, won't track results efficiently and won't make changes based on those results.

+ **Not adjusting bid amounts:** Just as the copy and keyword selections should be edited in order to minimize CPA costs, the bid amounts should be adjusted as the quality score increases. By carefully monitoring the advertising results, your PPC company should be making adjustments to bids to increase your ROI.

+ **Not differentiating between the Content Network and Search Network:** Now that you've learned about the nature of the Content Network vs. the Search Network, it's obvious that each needs their own campaign. An experienced PPC company should keep these two networks separately in order to keep your cost per acquisition low.

+ **Not using placement performance data:** If the Content Network is part of their strategy for your business, they should be analyzing the performance data and adjusting your campaigns to emphasize content networks that are producing more conversions for your company. You can easily check if this is the case by logging into your PPC account and looking at the campaign results.

QUESTIONS TO ASK ABOUT CPA MINIMIZATION

+ **How do you plan to reduce my excess ad spend?** Your potential PPC company should respond with a clear plan to adjust your spending in one of several areas – bidding strategy, keyword refinement, professional copywriting and search vs. content network. If you've been running your own PPC campaigns, there may be several simple tweaks that your PPC company can make initially to your accounts to instantly decrease your CPA. They should outline immediate steps, intermediary steps and long term strategies to get you the best ROI. Watch out for unclear answers or answers that imply that spending more money in PPC will lead to more sales. This isn't always the case and you should be wary of a company that doesn't see CPA minimization as a natural part of campaign management.

- **How do you plan on improving my Quality Score?** Even though your PPC company may be working with other PPC systems, they can't discount the importance of having an improved Quality Score within the Google Adwords system. A PPC company should help you establish your own high Quality Score Google Adwords account and not simply rely on their existing group account in order to get your ads more clicks. If they use their own account, you are left without a campaign if you ever part ways in the future. They should respond with specific methods of improving Quality score, like those outlined above.

- **Do you plan on using the Content Network for my PPC ads?** Be wary of a company that completely rejects the Content Network as a valuable means of traffic and conversions. While they may initially be focused on the Search Network for ads, the Content Network should be part of their long term strategy. Since the Content Network can make or break your budget, you should get their opinion of use of the Content Network with regards to your specific campaign needs.

Case Study #2: Light Bulbs Etc!

One of the first steps JumpFly took with this company's PPC account was to reduce spending in the Content network. They also created new Google Adwords and Yahoo Search Marketing accounts that were properly structured for the best results and the highest conversions possible.

Case Study #3: All Things Aquarium

Since All Things Aquarium was new to the world of PPC, the primary method of minimizing CPA was geographically targeting the ads. This way the PPC ads only appeared for local areas and not all over the country.

AUTOMATION

In your quest to find professional PPC services you may run across one of the many automated programs out there that claim to offer completely automated push button PPC services. This may seem tempting if you are aiming at keeping your PPC costs low. The thought of spending one low fee for a software program or service seems to be a viable option compared to the high cost of using a PPC management company.

However, PPC companies have insight and acumen that even the most sophisticated PPC automation tools cannot offer. At best, PPC automation can provide assistance with budgeting and bid management. But true increases in ROI can only be achieved if a capable PPC manager is using the automated tools.

An experienced PPC company will use automated PPC software to compliment their experience and their service it helps them effectively use their time so they can get to the core tasks of managing your account. Automated software programs allow them to focus on the tasks and responsibilities that cannot be done with a system, like copywriting and campaign analysis. Analysis tools help them minimize the waste of your time and your money. There are several different forms of automated tools that your PPC company can use.

BID MANAGEMENT SOFTWARE

This type of software will track your bids and adjust them many times a day so that your PPC are optimized to maximum efficiency. The software will manage your bid amounts and allow your PPC company to optimize them with many different metrics, such as cost per action or ROI. For more on bid management software options, see the bid management section of this book.

TRACKING SOFTWARE

Tracking software: Your PPC company will make use of tracking software (above and beyond the tracking capabilities available in each PPC program). This will help them set up customized tracking metrics that can inform their decisions with your PPC account. There are several different options available for PPC professionals. ClickTracks is reliable and allows the PPC company to use viewer traffic through many different visitor paths. Keyword Max, IndexTools, WebSideStory, Efficient Frontier or Omniture are also popular choices for PPC professionals. If you'd like to track your results yourself, ConversionRuler is an inexpensive and easy to use tracking program that is efficient for a cursory overview of your PPC account.

CLICK FRAUD MONITORING SOFTWARE

Fraud protection is important in pay per click bidding. Invalid clicks cost you money and if an individual or company is maliciously clicking on your links, your PPC budget will go through the roof. Major pay per click search engines offer fraud protection as an existing part of their service, however your PPC company should also be validating this on a larger scale.

Click fraud protection tools allow your PPC company to quickly and easily spot fraud. Who'sClickingWho software and Keyword Max Click Auditor are common tools for PPC companies to use. ClickTracks also includes a click fraud monitoring tools.

Your PPC company should have these tools already and be able to use them for your PPC campaigns. There should not be extra charges above and beyond your initial fees for tracking software. Professional PPC companies use automated tools as part of their management, and it should not be something that they have to buy in order to manage your account.

RED FLAGS FOR AUTOMATION

* **No automation used:** A PPC company that does not use automated solutions like those outlined above are not using their resources wisely, which means they are not using your resources wisely. Automated tools are not a way to "cheat" the PPC search engines. They are time savers and analysis tools that enrich a company's ability to manage your PPC account.

✦ **Using a poor quality automation tool:** By asking about their preferred automation tool (see below) you can do some research on the quality of the tool and see if it provides quality results. See the PPC Automation Software chapter of this book in order to learn more about evaluating this type of software.

QUESTIONS TO ASK ABOUT AUTOMATION

✓ **Do you use automation software?** The answer should be yes. Be wary of a PPC company that claims they do everything by hand, because their work may be inefficient.

✓ **What automation tools do you use to manage my PPC account? How do they help you make my PPC campaigns more profitable?** The PPC company's answer to this question will give you insight into their automation process and help you understand how automation will help them give you excellent service.

PPC CASE STUDIES – FINAL RESULTS

Throughout this chapter, we've referred to three different PPC accounts that were managed by professional PPC companies: Stein Diamonds by WPromote, Light Bulbs Etc! by Jumpfly and All Things Aquarium by Jumpfly. Each company used campaign analysis, keyword selection, effective ad creation, CPA minimization and automation in order to manage PPC accounts and increase ROI. Here are the results of their PPC efforts.

WPROMOTE'S CASE STUDY FOR STEIN DIAMONDS

After taking over the existing PPC account for Stein Diamonds, WPromote was able to lower the cost of Google Adwords spending by 73% in the first seven months. The cost per click decreased by 42% and the cost per lead decreased by 57%. With the reduction in spending, they would be able to bring in 46% more conversions if they had the same PPC budget as they did before WPromote began managing the account.

JUMPFLY'S CASE STUDY FOR LIGHT BULBS ETC!

After the first month of service from JumpFly, Light Bulbs Etc! had twice as many orders as they had in previous months while ad spend remained the same. The average cost per conversion was reduced by more than 50%. Over the next six months, the management requested that the budget be increased to accommodate for more exposure in their niche. This produced new business for less than half of the previous cost-per-conversion.

JUMPFLY'S ALL THINGS AQUARIUM CASE STUDY

Since All Things Aquarium was brand new to PPC, their results were immediately impressive. Within the first month, they received more than 200 qualified local visitors to their website, increase phone calls and several new clients. The success of their PPC campaigns allowed them to reduce spending in other areas, such as Yellow Pages advertising. They increased their budget to produce more than triple the website traffic and many new leads. The company surpassed their total revenue for 2006 by the first six months of 2007 with the help of JumpFly's PPC management.

CONCLUSION

PPC marketing is a detailed and complex form of advertising that can be very effective if it's applied with the right methods. When you are considering a PPC company for your advertising needs, you can apply the guidelines in this chapter to evaluate their level of competency. Educating yourself on the basics of PPC will help you identify whether or not the company is able to meet your needs.

CHAPTER FOUR:

LINK BUILDING COMPANIES

We briefly touched on the importance of link building in the SEO chapter, but now it is time to take a closer look at the steps a link building service should take on your behalf. Link building has long been an important part of building websites and, with the right link profile, your traffic and conversions can increase.

Link building is the process of accumulating links that point to your site and developing links to display on your site to other websites. It is a vital part of increasing your site's popularity because it is how search engines determine your site's worth and popularity. When your site's popularity increases, it will appear higher in the search engine rankings. Links will be an important part of your traffic building practices. This chapter will educate you on the correct link building practices that a quality company will use to increase your site's impact online.

THE IMPORTANCE OF BACKLINKS

If a reputable website is linking to your website, the search engine spiders count that as a "vote" in your favor. The more "votes" your website has for your specific keywords terms, the more likely your website will be found in the top pages of the search engine listings. This is an overly simplified way to look at the process because there are other factors involved in the search engine listings. However, it does get the basic concept across – backlinks are important for establishing your site's relevancy to particular keywords.

In addition to site relevancy, backlinking also helps your target market find you. If a popular blog in your niche links to your website, that link is exposing you to a whole new section of people who may not have actively sought you out. Backlinks can open your site up to being optimized for a variety of different keywords. Backlinks can make a website show up in the search engine results, even if that page is not optimized for that term.

Receiving links from other trusted sources makes your website a part of a community of sites on your topic. For any given topic, there will be a core of key sites that are relied upon for quality information or quality products. Although you don't necessarily want to link to your competitor and they don't want to link to you, chances are both of your sites will be linked to by a third party source to identify you both as leaders in the niche.

Getting links from reliable sites and becoming part of the well-known resources makes you part of the "inner circle" for your topic. Once other people link to you as a resource, you'll find your site will get additional links just because you're part of this circle. This behavior was noted as part of the way the web works back in 2000. AltaVista, Compaq and IBM put together a linking theory that is all the more accurate today because Google made links an important part of their algorithm (with other search engines following suit). The "Bow Tie Theory" shown below theorizes that all websites can be broken up into three different categories.

Approximately 30% of sites fit into the "Core" of the web – these sites have backlinks and are linking to other relevant sites. Your goal is to be in the "Core" because those websites receive inbound and outbound links. The search engine spiders see an exchange of links in both ways as more valuable than a site that only has links one way or, even worse, a site that has no links at all.

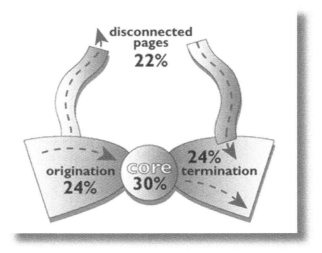

(source: http://www.almaden.ibm.com/almaden/webmap_press.html)

The left side of the diagram shows that 20% of sites have links to the core but very few links coming back. They are called "origination" sites. "Termination" sites represent 20% of the equation and have links coming from the core but none reciprocating. The remaining percentage of websites has no relationship with the core, but link to origination or termination sites or is independent.

Being part of the core of website traffic means that your website has more popularity and therefore it will have more importance in the eyes of the search engines. Search engines aren't sophisticated enough to be able to determine the quality of the content on a website. They can't "read" and determine whether the page is helpful or not. The keyword usage and backlinks are the data that they have to go by in order to determine a website's importance.

Being part of the core will also help you attract other links from independent websites. The techniques your link building company will use will create incoming traffic from a variety of different sources that will label your site as an authority in your niche area, or at least move you closer to being an authority in the niche. Keep in mind that back linking takes time to develop. The foundations that your link building company builds today will help support your website for years to come.

When Google became a popular search engine in the year 2000, links became even more important than ever. It was the first search engine that used popularity as a major part of their algorithm. This meant that links (popularity votes) became an essential part of how a website was ranked. Other search engines followed suit.

THE RIGHT AND WRONG WAY TO BUILD LINKS

Although getting links is not hard, it can be difficult to get qualified links. Qualified links are links that help build your site's authority. That is where a top-notch link building service can help your business. There are several link building services out there that completely ignore this basic link building principles. They sell their clients on packages of thousands of links and promise that they'll get a boost in the rankings. You might have gotten an inquiry from one of these link building services in the past. They often send spam to website owners and tried to sell their link building packages.

Although using this type of service will greatly increase the number of links to your website, it will harm your site in the long run. The quality of your inbound links helps the search engines determine how much quality your site has. If the inbound links are from link farms and other spam websites, your quality and therefore your search engine ranking will fall.

The most effective links in terms of building your traffic and your search engine ranking are natural links, or links that appear to be natural. Your link building service should seek to reflect natural linking as much as possible in order to keep your website in good standing with the search engines. Here are some of the most important characteristics.

o   Natural links mostly link to the inner pages of your website. This is attractive to search engines because it shows more specific and relevant.

o   Natural links come from different sources. This proves your relevance to the topic since you have links coming in from many different sources.

o   Natural links are built slowly. Link building that follows a slower pace reflects the natural process of link building.

o   Natural links are not reciprocal. This is done to prevent people from setting up schemes to exchange links and artificially boost up search engine rankings.

Natural link building can be done artificially if your link building service follows a few simple steps. In order to make sure that your link building service should mimic natural link building techniques as much as possible. Natural style link building will protect the integrity of your site and prevent search engine penalties. In this chapter, you'll learn about the correct methodology for building quality links.

GENERAL LINK BUILDING RED FLAGS

* **Building links too fast:** This is one of the biggest reasons that websites get flagged as "spam." In order to mimic natural link building, your link building company should adopt a long term strategy for link building. One hundred links overnight will set alarms off in the eyes of search engines. Your link building company should present you with a strategy that will pump up your link profile without being too overly aggressive.

* **Reciprocal linking:** This technique is not considered a viable way of obtaining links by the search engines. The main reason reciprocal linking is frowned upon by search engines is because it does not imply quality. It should be noted that reciprocal linking can be done on a very small scale, but sites that are large hubs that manage link trading and arrange swaps should be avoided.

* **Paid linking:** Just like reciprocal links, paid links are not a trustworthy way of obtaining links. By using paid links you aren't showing that you can obtain links the regular way. Make sure your link building company explicitly avoids paying for links. This includes subscribing to services that automate link building or arrange for link exchanges.

* **Over-optimized linking:** In order to appear natural to search engines, your links should use several different keywords as anchor text in order to get a varied amount of links coming into your website. Your link building service should use a variety of important keywords as anchor text.

♦ **Linking solely to the homepage:** Your links should be coming into internal pages of your website as well as the homepage. This will help build the authority of your site and it will appear more natural to the search engines.

♦ **Focusing linking efforts only on incoming links:** Incoming links are only part of an effective link building strategy. Without changes to the content on your website, you won't get natural backlinks from other websites. Without linking out to other sites, your website's link building efforts will appear unnatural to the search engines. Your link building company must present you with a comprehensive plan for tackling all three of these key areas.

GENERAL QUESTIONS TO ASK ABOUT LINK BUILDING

✓ **What is your link building methodology? Is it automated?** Successful link building companies will be interested in getting quality back links for your website. Finding quality sites to link to can take a long time, so some link building companies have tried to automate the process. Automation should be used with caution because it can often lead to reciprocal linking schemes. One such scheme is to send out automated messages to other webmasters in order to try to obtain links. Often these messages are mass sent to anyone with a website, regardless of their niche or website standing. This can lead to a penalization from Google. A good link building company will concentrate on finding natural style quality links.

✓ **What is natural linking and how does your company replicate natural linking?** This tests their knowledge of natural linking and will help you evaluate their methods of link building.

✓ **How soon can I expect to see results from link building?** A quality link building company should explain to you that, like other SEO steps, the effects of link building may not be seen for weeks or months. It is a cumulative process that you'll see the effects of long term once they set the wheels in motion. Steer clear of companies that promise instant results.

THE BASIC COMPONENTS OF QUALITY LINK BUILDING

Links are the basic connector between sites and the easiest way for search engines to determine relationships. Even as search engines become more refined with their ranking techniques, back links will still be an important part of ranking. These areas represent the basic method that your link building company should use to create a complete link building campaign. Depending on your budget, you may use all or some of these services.

ASSESSMENT

The first step any quality link building service should take is to conduct a complete analysis of the current links that are coming into your website. By determining the current standing of your links, your link building company can set measurable goals for their link building campaign. Your link building company will begin determining how many links you currently have and what their quality is. This will give them a starting point to work with and help them refine the linking strategy for your specific case.

There are several different tools your link building company can use to determine your existing backlinks, and you can use these tools as well to check on your site. You can use Google and enter the following search query:

link:www.yourwebaddress.com

For Bing use the following search query:

"yourwebaddress.com" - site:yourwebaddress.com

To check backlinks in Yahoo, type the following into the search engine:

http://www.yourwebaddress.com

Alternatively, you can visit http://www.linkpopularity.com and obtain links to each of these reports. Please note that this site lists Bing by its former name "MSN search." Here's an example of the Bing results for Yoga Direct.

(source: http://www.bing.com/search?q=%22yogadirect.com%22+-site%3Ayogadirect. com&FORM=MSNH)

The highlighted area shows you how many websites are linking back to YogaDirect.com according to Bing. Here are the results from Google and Yahoo.

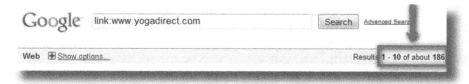

(source: http://www.google.com/search?hl=en&q=link%3A+www.yogadirect. com&sourceid=navclient-ff&rlz=1B3GGGL_enUS324US324&ie=UTF-8)

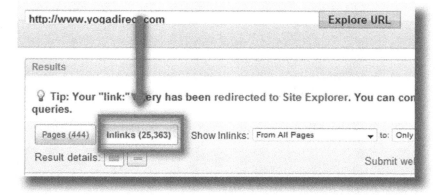

(source: http://search.yahoo.com/search?p=http%3A%2F%2Fwww.yogadirect. com&toggle=1&cop=mss&ei=UTF-8&fr=yfp-t-701)

As you can see, there is a quite a bit of difference between the search engines. They use different ways of evaluating backlinks so it's important that your link building company view each search engine when they determine how many backlinks you are receiving. It's also important to keep this in mind if you are checking up on your link building company's progress. Be sure to check all of the search engines so you get the entire picture.

After evaluating your backlinks, the link building company will determine if you need additional backlinks from high quality sources Other options can include getting rid of badly placed links that point back to your site. Your company will evaluate the quality of the links you are getting based on their PageRank and a variety of other factors.

In addition to checking which sites are currently linking to your site, your link building company should look at your competition's websites using the same tools. By looking at the competitor websites, they'll be able to see what your website is up against in terms of backlinks. This will help establish a benchmark for the quality and amount of links that your website should have. Secondly, your link building company will be able to create a list of sites that should be linking to your site. If the sites are linking to your competitor, there's a good chance that they will link to you as well (as long as they are third party sites and not part of your competitor's network of site).

From these evaluations, your link building company will be able to determine which sites could be linking to you. They will develop a link building plan based on the quality and number of links that are available in your market.

RED FLAGS FOR NEEDS ANALYSIS

+ **No needs analysis or forcing a boilerplate link building method**: An experienced link building company will know what works and what doesn't. Even if they have their link building methods down to a science, they will always take analysis of your site's links and determine how to customize that strategy to your needs.

+ **Not considering the backlinks from competing sites**: Competing sites are essential in the evaluation and planning of a backlink building campaign. Your link building company should ask about your major competitors and evaluate their backlinks.

QUESTIONS TO ASK ABOUT NEEDS ANALYSIS

✓ **How will the backlinks of my competitors be used in the backlink strategy of my website?** There are many correct answers to this question. As long as the company indicates that they will be taking your competition into consideration, they are on the right track.

✓ **How will the backlink building campaign be structured based on my company's needs?** This question will reveal whether or not they use boilerplate methods or truly customized link building strategies.

DEVELOPING A LINK BUILDING STRATEGY

A link building strategy is a methodology of building backlinks in a way that will bring in new customers and will increase the ranking of the website in the search engine results. Once your link building company has evaluated your needs and determines the status of your incoming and outbound links, they will develop a link building strategy.

A link building strategy is not a magic pill for traffic and increased ranking, and your link building company should communicate to you that proper link building takes time. Initially, they'll be setting up the foundation for your backlinks and then monitoring how your site is being linked to organically.

The basic steps for a link building strategy are follows:

1. Create valuable content.

2. Link to other valuable content.

3. Ask for links quality sites that are relevant to your topic.

That is the process in a nutshell, but there are nuances to each step that will make the difference between having a successful link building campaign and a campaign that harms the integrity of your site. The first point to note is that while the number of links you have is important, the quality of those links is more important. Link building for link buildings sake will get your site nowhere. Since the search engine algorithms operate by popularity, the quality of the sites linking to you matters.

Earlier we referred to link credibility being passed on from site to site. Each search engine uses a different system for determining popularity. Google uses a system called PageRank, which is a score on a scale of 1 to 10 for a site's value. The higher the PageRank number of a particular page, the more important it is in the eyes of Google. The important thing to note about PageRank is that it is passed along from page to page. If your homepage is ranked at a PR5 and it links to your internal pages, it "passes along" some of that PageRank. This is sometimes referred to as "link juice". Your other pages become more popular by being associated with your popular page.

The same goes for external links you receive from other sites. A half a dozen links from high PageRank websites will help your website more than hundreds of links from PR0 or PR1 sites. You can also increase your site's popularity by linking to other quality sites related to your topic. If your link building company does not include content or outbound links as part of their strategy, there is a problem. All three components are necessary in order to build long term links that will bring traffic and increased search engine ranking to your website. We'll look at each of these areas independently so you can understand how they work together.

QUALITY CONTENT

We touched on the concept of quality content in the SEO chapter where we shared the difference between keyword optimized text and keyword stuffed text. But the use of keywords is not the be all and end all of determining what makes good content. Quality content is important because it encourages backlinking. Other sites will link to your site if you have something important to share. The content must be relevant to your topic and it must be "link worthy." In this section, we'll go over a few key concepts for quality content and what your link building company should be creating for your website.

If your link building company tells you that you can build quality backlinks without changing the content on your website, they aren't using the right methodology. The end result of a backlinking campaign is that a potential customer arrives at your site (through the search engine rankings or through directly clicking on the link from another site).

Your link building company will create unique content that relates to your niche and your services on your website. They can create high quality tools for you that people will link back to. They can create studies, lists and other forms of content to encourage links. If you decide to use blogging as part of your search engine optimization strategy, your content will be updated frequently with "link bait" type posts. However even if your website will be static (not updated with content on a daily or weekly basis), there are plenty of opportunities for your link building company to create content in order to encourage links.

Developing the quality of content on your website makes your website easy to link to. Quality websites will link to you and spread your message if you have something valuable to say. Your link building company or content creation company should improve your content so it attractive to the general public and other website owners.

If you have a service based business, you can publish tips articles, how to articles or general advice on your business. For example, a criminal defense lawyer could create articles detailing a defendant's rights and what to do if you get arrested. A life coach's website might include articles such as "5 Steps to Break through Old Patterns" or "How Do You Know if Working with a Life Coach is Right for You?"

Product based companies can host content that explains how to solve common problems with the products they sell. They can have content created on the benefits of particular products or how to choose the right product. As long as the articles are helpful without having an overly promotional tone, they will invite links.

Creating lists for your website is another excellent way to prepare content for link building. Lists are versatile because they can be used in any niche to create content that is worth linking back to. For example, your site could list the top ten innovations in the industry in the last ten years. You could use a list of "10 Easy Tips for...." or "Top 10 Myths in...." For a closer look at content creation services, please see the Content Creation chapter of this book.

RED FLAGS FOR QUALITY CONTENT

+ **Not adjusting your content for link building:** Your link building company should suggest content practices that will help your credibility in the niche and encourage backlinks.

+ **Use of "spammy" content:** Content should be geared toward the niche and include some keywords, but it shouldn't be keyword stuffed (see the SEO chapter or Content Creation chapter for examples of keyword stuffed content).

QUESTIONS TO ASK ABOUT QUALITY CONTENT

✓ **What content changes should I make to my website in order to encourage links? Will you make these changes for me?** Your link building company should outline the content related steps that they'll take in order to help encourage links. If the company says that they don't deal with content, they aren't a full service link building company.

✓ **Are you familiar with link bait articles? What type of link bait would you suggest for my website?** A qualified company should be able to tell you what link bait is in a simple sentence. Give preference to companies that can come up with interesting link bait ideas or have link bait experience in your niche.

INBOUND LINKS

After ensuring that the content on your website is in good order, your link building company will begin to bring in links from trusted sites. This is the best way to get more "link juice" to your site and increase its importance. There are several different strategies that your link building company can use in order to obtain quality backlinks. Before we get into the specific techniques, you need to be aware of two methods that should be avoided: purchasing links and obtaining links from link farms.

There are marketplaces where webmasters can purchase links from one another. They offer a spot on their website in exchange for a monthly or flat fee. Generally, these links aren't great quality and to top it off the practice is discouraged by search engines. In Google's Webmaster tools' section on Link Building Schemes it defines "buying or selling links that pass PageRank" as a scheme. Your site could be penalized if your link building company buys links. In fact, Google has a location on their website where they ask people to report link buying websites.

Almost as bad as link buying is making your site's listing part of a "link farm." A link farm is a website that makes a practice of linking to anything and everything. What results is a large un-targeted website with thousands of links which won't help your ranking and it won't help your traffic. There are legitimate directories out there that will list your website; however there is a distinction between a directory and a link farm.

Link farms will:

o   Not have any connection to the subject matter of your website.

o   Show "bare" links with no description or helpful information for the website visitor.

o   Send you an unsolicited email asking for a backlink.

o   Ask you for a link from your links page, even if you don't have one.

It's essential that you ask your link building company where those links are coming from and monitor any links you have coming into your website. These two practices can damage your site's ranking in the long run. Fortunately, there are several perfectly legal ways to obtain incoming links that will help your site increase in search engine ranking and optimize your site.

ARTICLE DISTRIBUTION

Your link building company can use article distribution to create backlinks to your site from high quality article directories. Submitting articles that are directly related to the content of your site can help drive traffic and increase your search engine rankings.

Article directories are large websites that accept articles on a wide variety of topics. They allow authors to add links back to their website in the resource box. In normal situations this box will give details on the author or how to get more information about the topic. But an expert link building company can turn this tool into a powerful method for backlinking.

By selecting the anchor text used in the resource box your link building company can increase your site's search engine rankings for those keywords. As an added bonus, you may also get additional traffic from the article distribution. Articles from directories can be republished on other sites, as long as the resource box is reprinted as well. In theory, your article could go viral over the Internet and bring hundreds of quality backlinks to your site.

Make sure your link building company is submitting to valuable directories, from a PageRank standpoint. Quality over quantity matters the most with article marketing. EzineArticles (http://www.ezinearticles.com), GoArticles (http://www.goarticles.com) and IdeaMarketers (http://www.ideamarketers.com) are among the best in terms of quality and traffic. The article directories should have some submission standards. If the directory doesn't have submission standards (post length, amount of backlinks, quality of content) your article won't be in good company.

The quality of your articles is important in this method of building backlinks. Your link building company can create these for you, or you can hire out content creation to a different firm and have your link building company place these articles. For more information on obtaining content for article submission, see the content creation section of this book.

ONLINE PRESS RELEASE DISTRIBUTION

Professional media use press releases all the time to distribute important news about a company or a person. Online press release distribution has revitalized this form of communication and is very helpful for building backlinks. With a well placed press release, you can expect a boost in traffic to your website. Your press release is distributed to press release distribution services. These services will push the press release on to news outlets around the world. Sites like PRWeb.com and PRLog.com will help increase the distribution and traffic to your website.

When your link building company uses press release distribution, your press release may be distributed to hundreds or thousands of websites. Each of these websites will contain a back link to your website. By using anchor keywords in your press release the link building company can give you a backlink and a link building boost. You'll also obtain listings in the "news" sections of the Google and Yahoo search engine listings. This can help your site build credibility.

Although press releases are easier to distribute these days, it does not mean that your link building company can create a press release out of thin air. The press release should contain newsworthy information. Overly promotional press releases are less likely to be picked up and distributed. For example, a press release with the title "Daylen Golf Accessories Updates Website" will be passed by.

Alert your link building company when you have a new project on the horizon. Your business expansion plans can be dovetailed into a new promotion and back linking opportunity. Details on press releases and press release distribution can be found in the press release section of this book.

FORUM AND BLOG COMMENTING

Link building services can use the power of forums and blogs to create valuable links to your website. Forums and blogs with high page rank that are closely related to the topic of your site are excellent sources for backlinks. They are fairly simple for your link building company to obtain and they can be a source of targeted links and targeted traffic.

Forums are websites that are created to allow people who share interests to communicate through forum posts. Messages are shared between all of the users and any links present in the messages or in the signature files of the forums posts will be cataloged by search engines.

Blogs are websites that are hosted by an individual. The individual will post messages with varying frequency. Very active blogs have daily posts but others may only post weekly or monthly. Most blog owners allow comments on their posts. These comments have space for links back to your website.

Before your link building company begins distributing any blog comments or forum comments on your behalf, they need to make sure that those sites have "do follow" links. If a site is using "no follow" tags in their HTML it will prevent any link benefits from coming your way. Posting on blogs and forums that use "no follow" tags are a waste of your time and your money.

Your link building company should find forums that are directly related to your niche. They should evaluate the forums and determine which ones are most active. Many forums are virtual ghost towns with no recent posts or very little user activity. Even though the primary purpose of obtaining backlinks is to increase search engine ranking, it doesn't hurt to have some additional traffic come in from the users of the forum. If the forum is inactive, this just won't be possible.

Your link building company should register an account and then set up a profile that contains a link to your website in the signature file. The signature file is displayed each time that user name makes a post. This way each post will contain a link back to your website. Your link building team should take care in the kind of posts they'll be using in forums. Many forums have anti-spam rules and will ban an account if they suspect that it is being used solely for advertising purposes. Your link building team should use helpful and informative posts in its forum posting efforts.

Blog owners will closely monitor comments to be sure that they aren't spam. Your link building team should take care to post helpful comments that are directly related to the post and work in a mention to your website. They can use tools like the DoFollow Diver tool at InlineSEO in order to find do follow blogs that will allow links.

If your link building company does not offer forum posting or blog commenting services, you can find companies that specialize in these forms of backlinks. You can also locate freelance content providers that will offer these services. For more on hiring content providers, see the content creation section of this book.

SOCIAL BOOKMARKING

Social bookmarking has emerged as an excellent tool for obtaining backlinks in the last few years. Social bookmarking sites allow users to share links that they find important or interesting with others. When a user signs up for a social bookmarking site, they can list sites that they like and appreciate. All of these sites receive a back link from the social bookmarking site itself. Users make "friends" on the sites and their favorite sites lists are shared with those friends. The friends have an opportunity to bookmark those sites as well.

Link building companies can use these sites as an opportunity to bookmark your sites and increase link power. There are hundreds of social bookmarking sites, but your link building company should focus on only the most popular sites. Digg (http://www.digg.com), Delicious (http://del.icio.us) and StumbleUpon (http://www.stumbleupon.com) are a few of the most powerful.

In addition to these general social bookmarking sites, there are hundreds of others that are more niche specific. Your link building service may be able to find social bookmarking sites that directly relate to your industry. If your link building service doesn't offer this specific service, you can find social bookmarking services from other companies or independent contractors.

Social bookmarking is part of a large category of Internet marketing called social marketing. New technology in the last several years has made the Internet more social and smart companies take advantage of these opportunities to marketing their business online. For more on hiring qualified social marketing experts, see the social marketing chapter of this book.

LINK WHEELS AND WEB 2.0 PROPERTIES

Web 2.0 sites, of which social bookmarking sites are a part of, have changed the way people interact on the Internet. LINK BUILDING professionals have discovered that Web 2.0 sites are an excellent way to build backlinks to a website. Web 2.0 sites are free sites that can contain written content, pictures, videos and, most importantly, a link back to your website.

Web 2.0 sites can be linked together in a link wheel to drive traffic and produce back links to your site. The concept of link wheels is relatively new. You may also see this same principle marketed as feeder site marketing or Web 2.0 marketing. Link wheels get their name from their basic set up. The center of the wheel is your website. All of the spokes of the other Web 2.0 sites will point back at your website. They will also link to one other Web 2.0 site.

Your link building company should choose six to eight Web 2.0 sites that have a high page rank. WordPress, Blogger, Squidoo, Yola, Hubpages and Weebly are some of the sites your link building company could use. Each of these spokes should be optimized for a specific keyword phrase that is related to your main site. Unique articles are put on each site and the sites are updated regularly, if need be.

Your link building company can expand your wheel by increasing the number of web 2.0 sites involved in the wheel. Your link building company can also combine social bookmarks with the link wheel in order to drive more traffic to any of the spokes which will, in turn, drive traffic to your website. Link wheels are a relatively new phenomenon for link building, so if your link building company does not offer it. You can request this service or obtain it from an independent contractor.

REQUESTING BACKLINKS

In the needs analysis step, your link building company should identify sites that are linking to your competitors' sites but not yours. These are prime targets for link building. Your link building company should have a clear idea of why each of these potential sites would want to link to your site. They should also evaluate and determine how those sites will work into your overall link building plan.

Typically, your link building service will send a message to the webmaster of the site from which they'd like to obtain a link. Each site should receive its own message and the message should be customized to that website. If your link building service sends a boilerplate message to the potential linking websites it will greatly reduce your ability to get links back to your site.

In some cases, your link building company may request access to a unique email address from your website so they can request links from your domain. This can increase the chances of people responding since the request appears to be coming from your company, and not a link building company. If the link building company passes on the letter to you for approval, make sure it includes:

+ A named contact, if one is available.

+ The address of the page on the site to which your site would like to link.

+ A summary of the benefits of their visitors coming to your site.

+ A linking code.

+ A request to get in touch.

Your linking company should keep track of these link building attempts and continue to send out requests. They will monitor to make sure that the links are coming from internal pages that are related to your site's topic.

ADDITIONAL LINK BUILDING METHODS

Do not be surprised if your link building company suggests some of the following ideas in your link building plan. Link building plans are customized to the needs of your website and your link building company may employ any one of these techniques to increase your backlinks.

o Adding links from their own network: Many professional link building companies have their own aged domains that they can use to provide quality backlinks to your website. If they've taken this step in creating domains for link building, they will probably have several domains that are targeted for specific topics.

o Adding a blog to your website: Blogs make it easier to post link bait content and encourages people to use social bookmarking to create additional backlinks.

o Hosting a contest: Depending on your market, this can be a viable way to build backlinks. Giving a discount on a product or service, or simply giving away a free prize like a book on your website will create a great deal of backlinks from other people in your niche and also websites that catalog free prizes.

o Adding reviews to related products on other websites: Your link building company can use Amazon.com and other retail sites in order to provide additional backlinks. Posting reviews is free and these sites normally have high page rank. By creating a review, your link building company will not only establish a link but may drive traffic from the retail site to your site.

RED FLAGS FOR INCOMING LINKS

- **Too many links, too quickly**: Link building should not be done "blast style" where thousands of links are distributed within a short period of time. The more natural the links appear in the eyes of the search engine, the better.

- **Relying too much on one method of link building:** The key to appearing natural is in the variety of links. If all of the links to your website come from the same site (like Ezinearticles.com), it will look suspicious to the search engines.

- **Submitting articles of little value in order to only get backlinks:** Article distribution is most effective when the articles offer some value to the reader. Most quality article directories have editors that screen out poorly written or poorly structured articles. In addition to creating backlinks, the articles can also bring in traffic from readers so it's important that the articles being distributed on your behalf are helpful to people in the niche.

- **Distributing multiple press releases regardless of newsworthy importance:** Press release distribution is only effective if the content is actually newsworthy. If your link building company encourages the use of press releases without asking for newsworthy items or events they can promote, they are abusing press release distribution websites.

- **Spamming forums and blogs with comments:** If your link building company will be using these forum marketing and blog commenting to build links, they need to go about it in the right way. The quality of the comments and the timing of them should be done is such a way that they won't set off any spam flags in the eyes of the website owners or the search engines.

The comments should be related to the topic being discussed and shouldn't be overly promotional.

+ **Building low quality spokes for link wheels:** Link wheels are most effective when they can function as standalone resources for your niche. While they should have links back to your website, they shouldn't be overly promotional in nature. Your link building company should make the content and design of the "spoke" sites of good quality.

+ **Automated link exchange emails:** If your link building company sends out link requests on your behalf, they should be personalized to the website and website owner to which they are going. With no personal names and no personalization, their efforts (and your money) will be wasted.

QUESTIONS TO ASK ABOUT INCOMING LINKS

✓ **What is your timeline for building incoming links?** This will let you know if they have the right type of link building style in mind (slow and natural-looking). If they respond with no timeline or claim they can get you thousands of links in a month or two, look for service elsewhere.

✓ **Will you use an email address that I provide in order to submit articles (or press releases, forums, blog comments, social bookmarking, Web 2.0 site creation, etc)?** This is important because you want to maintain control over these accounts if you part ways with the link building company. It's essential that you can login to these accounts and use them in the future. If they use one of your email addresses you can also monitor the type of content they are using on your behalf to make sure that it is quality.

OUTBOUND LINKS

Outbound links are the final piece of the backlinking strategy. Most new website owners would assume that linking out to other sites will draw people away from their website. On the contrary, linking to other websites can help make you part of the "core" of websites that we showed earlier in this section.

The websites in the core of that diagram are the known authorities on the topic. They have links coming to them and also have links going out to other sites. It is theorized that outbound links are a component of the algorithm that search engines use to calculate a site's popularity. By linking out to other websites you can improve your popularity for a term and become a resource for your target area. However, the outbound links need to be done in a specific way so as not to look suspicious and get your site flagged by the search engine spiders.

Watch for these rules when your link building company is implementing an outbound links plan.

o   You should have some outbound links, but not too many. Since there is no exact way of knowing which number search engines have deemed "too many" you'll have to rely on your link building company's expertise on this one. It's safe to say, though, if your link building company wants to turn your website into a link farm or populate a page with irrelevant links, it's not a good sign.

o   All links should be relevant. All of the outbound links on your website should be related to your website's intent. World Class Chocolates would link to a website where they sell handmade gift baskets, or a perfume site or a stationary web store. These are all under the "gifts" umbrella. It would not link to a machine parts website or a pregnancy website, no matter how popular those websites are.

o   The links should use appropriate anchor text that relates to the nature of the website, and also related to your topic. Outbound link anchor text helps improve your website's pagerank for those terms. This means that if World Class Chocolates links to a gift basket website with the words "customized gift baskets" it may rank for that term, even if customized gift baskets appear nowhere else on that site. Although the web searcher will not find customized gift baskets on the World Class Chocolates site, they are related enough that the customer may change their mind and explore the site to make a purchase. At the very least, they've been exposed to the website and may bookmark it for return.

o   Your outbound links should be to high ranking sites. If you have a choice between linking to a PR2 site for your niche and a PR4 site for your niche, go for the latter.

Outbound links are essential to putting your site in a place where other people will want to link back to you. If your site is brand new, you're not going to receive many inbound links from other website owners until you prove your link worthiness. Establishing a smart list of popular outbound links with the right anchor text will make your site more worthy of inbound links in the future.

RED FLAGS FOR OUTGOING LINKS

*   **Labeling a page on your website "Links":** Nothing looks more suspicious to a search engine than a long list of links on a page titled "links." Your link building company should structure your outgoing links in such a way so that they appear as a natural possible. They can be incorporated to blog posts, listed in a resources article or distributed on different pages. Your link building company should take steps to frame outgoing links in the right way.

*   **Linking to the same types of sites or linking to unrelated sites:** Both of these tactics are examples of the wrong way to create outgoing links. Your link building company should create a diverse outgoing link profile, but those links should all be related to your site's topic. For example, the links from a financial planner's website should go to financial related websites, like estate planning information, credit card debt reduction information, places to get free credit reports, etc. The links should not all point to high yield IRA websites, or any other single topic, just because you want the site to be optimized for that topic.

- Using the same anchor text again and again: This looks unnatural to search engines and will put up a red flag for their spam filters. The anchor text for the outgoing links should be related to your website, but they should also be different from one another. Monitor your website to make sure that your link building company is using the correct anchor text.

QUESTIONS TO ASK ABOUT OUTGOING LINKS

✓ Is building outgoing links a part of your link building strategy? The answer should be yes. If they respond no, ask why not. It may be due to your campaign budget but it should at least be a consideration.

✓ How will you determine which links to display on my website? The answer should be that they focus on relevance to your website and overall important in the niche. Outgoing links should not be selected based on their likelihood that they will link back to your site. It may be an added benefit of linking to another site, but it's not the sole purpose.

RESULTS MONITORING AND REPORTING

Your link building company should be monitoring the results of their link building efforts in order to determine whether or not those efforts are paying off. Every company does results monitoring a little differently, but there are a few methods that are preferred. After taking the baseline statistics at the beginning of the process, they should follow up on an ongoing basis to determine how their steps are affecting one of many factors.

Here are some of the factors they may take into consideration as they are monitoring your backlinks and outgoing links:

o Link popularity – This will be represented in the number of your backlinks and an increase in your search engine results placement.

o Your performance vs. the performance of your competitors – This figure will be determined by your link company's ongoing

135

evaluation of your search engine listing and the search engine listings of your competitors.

o   New sites that link to you after link requests

o   News sites that link to your site without being asked

o   Evaluation of your traffic compared with new links on the site

Your link building company will likely monitor these metrics on a daily basis and then submit a formal report for your review each month, as long as you are paying for their services.

RED FLAGS FOR RESULTS MONITORING AND REPORTING

+   **Not evaluating comprehensive results:** If your link building company is simply looking at the number of links you are receiving and is not incorporating your traffic and conversion goals, they may be focusing on building massive links instead of the right type of links.

+   **Not providing you with regular updates:** Your link building company should stay in consistent communication with you about your link building project, particularly when they first take you on as a client. They should be working closely with you to determine content changes and source link bait ideas, press release ideas and ask about your goals for the campaign. After that, there should be monthly updates on the status of their efforts.

QUESTIONS TO ASK ABOUT RESULTS MONITORING AND REPORTING

✓   **How often will I be updated on the status of your link building efforts?** They should be updating you at least once a month.

✓ **How will you be gathering data on the success of the link building process?** They should be looking at search engine results for the keywords that your site is being optimized for, evaluating competing sites, keeping records on articles, press releases, social bookmarking and other submissions they've made on your behalf, and general increase in traffic and conversions.

CONCLUSION

Link building is a dynamic and changing form of optimization. You should review the strategies of the link building company and compare them to the strategies suggested in this chapter. Link building can make or break your website's rankings, so it's essential that you ask about their methods, follow up on their actions and track the impact that their link building efforts have on your website.

CHAPTER FIVE:

SOCIAL MEDIA OPTIMIZATION

In the last five years the world of social media has exploded as more and more people join social networking sites online and more companies are experimenting with social media tools. A social media optimization company will tackle the changing field of social media. It will help your company embrace Web 2.0 marketing and extend your traditional online marketing into new areas. Social media marketing can help increase your company's online reach and brand awareness, increase your reputation, develop valuable backlinks and improve your search engine results performance. Most importantly they will help your company meet the customers where they are – at social media meeting places.

Many business owners wonder what the difference is between social media optimization and SEO. While SEO focuses on how to improve search engine ranking by attacking how search engines view a site, social media optimization will improve how groups of web users view a site, your products and your services. Please keep in mind that if you already have PPC and SEO work done on your site, using social media is not going to replace those actions. Social media optimization (SMO) should compliment your other online efforts.

Social media marketing is by no means fast and it's a little bit different for every company that uses it. That's why it's so important have the help from an SMO professional who has the knowledge to be able to efficiently use social media to increase the profile of a company online. The goal of an SMO company will be to:

- show your company how to listen to the market through social media tools

- teach your company how to use those social media tools to communicate with your market

- spread the message of your business and your brand.

In this chapter, we'll go over the basic principles of social media and how an experienced SMO company can help you reach your audience in new ways. In addition to these criteria, topseos.com evaluates SMO companies on a monthly basis. You can see the latest ratings for SMO companies at http://www.topseos.com/rankings-of-best-social-media-optimization-companies.

WHY SOCIAL MEDIA?

Why should a business invest in SMO? It's a valid question considering how much of a marketing budget can be spent on SEO and PPC services, the two cornerstones of marketing online. Simply put, the Internet is changing and has been changing for the last several years. The techniques that once put a company on the map are falling out of practice.

Consider the following (excerpted from SocialNomics.Net):

o 78% of consumers trust peer recommendations, while only 14% trust advertisements.

o The #2 largest search engine in the world is YouTube.

o Wikipedia has over 13 million articles.

o There are over 200,000,000 blogs.

o Over half of all bloggers post content or tweet on a daily basis.

o   Word of mouth travels so fast in social media the term "world of mouth" is now frequently used.

o   More than 1.5 million pieces of content are shared on Facebook on a daily basis.

It is clear that social media sites (Wikipedia, Twitter, Facebook, blogs) are all popular and they offer your company a platform from which to present your marketing message in a low resistance method. By employing the services of an experienced SMO company, you can tap into the massive force of social media to increase your conversions.

In the world of social media, people are already talking about your industry and they may even be talking about your company. By using Google's Blogsearch or Technorati you can find what people are already saying about your company on blogs. Those conversations are highly visible and have the potential of being influential to other people in your market.

Without having a presence in social media, you are allowing other people to do the talking for you. If your company is not represented in the social media, you are not part of the conversation at all. Social media can help you stake your claim and accurately represent your company, your products or your service. It also gives you an outlet to listen to your existing customer base, which can help inform your company's future decisions. Social media accounts let you connect directly with customers instead of relying solely on market research reports and other disconnected forms of research. Want to know what people think of an idea? Ask them on Twitter. Need to know what your customers are looking for? Post a poll on your blog. It doesn't get any easier.

Most importantly, using social media can be a growing process. You can start small with one account on one site and build up from there. The first stage may be developing a blog that you use to create unique content for your industry, and then paying attention to the comments that you are receiving. As you start seeing the effects of social media campaigns, you can expand your reach to other sites and create a more expanded social media profile.

Most of the sites used in social media are free, so you are paying an SMO company for their time and expertise. Hiring a SMO company is the best solution because they know the ins and outs of handling social media and using it for the best effect. They can work with your staff so that your entire company is well versed in social media tools and techniques. Social media has a long memory and mistakes made in your initial days can haunt you in years to come. Handing over social media management to a skilled professional can help you get the most out of the marketing method.

SOCIAL MEDIA BASICS

The primary difference between standard marketing models and SMO is that the weight of the marketing message comes from groups of people rather than the company itself. A skilled SMO company will be able to show you how to influence the right people who will then go on to influence other people within that target market.

There are a growing number of social media websites that allow members of your target market to interact and build collective meaning. An SMO company will help you place your business, products and services at the core of those collectives. It's essentially meeting your market where they are and influencing others to sing your praises.

Social media has the following characteristics:

o **Participation:** Social media, unlike traditional one way marketing, encourages feedback and conversation between all parties involved. There is no longer a top-down hierarchy to communication.

o **Openness:** Social media tools are open for all participants to make contributions (i.e.: through voting, comments, or adding their own content). Participants don't have to gain special access to add content.

o **Conversation:** Rather than "broadcasting" a marketing message, social media allows companies to interact with customers in a new way.

o **Community:** Social media tools create communities in which all members have common interests or a singular common interest.

(Mayfield)

SMO can be expressed in two different ways, which we'll be discussing the following sections – strategy and tools. An SMO company will create a strategy for a company's online presence and then suggest tools to use in that strategy. They will create content ideas, production schedules and other marketing tasks (like contests, events and public relations opportunities). An experienced SMO company or provider will have a track record in developing strategies and using these tools. They will also be on the lookout for emerging tools that will help a company establish themselves through social media.

Common tool categories used by SMO companies today include:

+   Blogs (normally hosted on your company's website)

+   Microblogging (typically through Twitter.com)

+   RSS (syndication of blog and website content through 3rd party services)

+   Photosharing (typically through Flickr.com)

+   Podcasts (audio recordings distributed through your site and podcast directories)

+   Social bookmarking (using StumbleUpon.com, Digg.com, Sphinn.com and several other sites)

+   Social networks (using Facebook.com, MySpace.com and several other sites)

+   Forums (one of the first forms of social media where users can meet together and post messages)

+   Content communities (differing from forums because members upload content of their own to share, i.e.: YouTube.com or Flickr.com)

+   Wikis (either Wikipedia.org or a hosted wiki that allows users to create and edit content)

These general categories may all be used by your social media company, or the company may pick and choose depending on your needs. There are always new tools within each category that come onto a SMO company's radar which may be used by the company to achieve your results.

Speaking of results, it's important to note that like SEO, SMO is not an overnight process. By hiring an SMO company you are dedicating your business to trying out their methods for several months to a year. Many businesses falsely look for an SMO company to help create an overnight buzz about their product or service, and it just doesn't work that way.

SMO experts come in a few different varieties. There are large firms that house SMO departments or specialists that compliment their existing online marketing services. The benefit of working with a company such as this is that their broad experience can help you tackle other parts of the online marketing puzzle, if the need arises for different types of management. Large companies may also be more reliable in terms of communication and results.

Mid-range firms may be existing marketing firms that taking on social media responsibilities, or their also start up firms that specialize specifically in social media. Social media is becoming a field of interest for many internet marketers and some of them have created exclusive SMO firms that cater specifically to the social media needs of clients.

Finally, there's the stand alone SMO professional who offers social media services as a solo entrepreneur. With a specialist such as this, you'll get personal attention from someone who has "walked the walk" and made social media a part of their life. However, you may be stuck if the professional has too many clients and becomes too busy.

There are advantages and disadvantages to each level of service, but no matter what size you choose, you need to be sure that their skills and experience is a good match for what your business needs. The following red flags and questions will help you weed out the professional businesses from the people who have simply hung out a virtual shingle with the word "social media expert" on it.

GENERAL SMO RED FLAGS

+ **Claims that they are social media experts:** This may seem like a counter-intuitive red flag. After all, don't you want to have an expert in charge of your social media? However, social media professionals agree that social media itself is changing too rapidly for anyone to call themselves an expert. Good social media consultants think of themselves as guides to world of social media. If a company or consultant uses the term "social media expert" tread carefully and do your due diligence. (Brogan)

+ **Lack of a social media profile:** This is one of the biggest signs that a web development team, marketing team, Internet marketing scam artist or independent freelancer is attempting to expand to social media without having experience. Your SMO company should have strong profiles on Facebook, Twitter and other popular social media sites. They should have a blog that is active, with frequent posts and comments from other visitors. They should get links to their blog from other respected sites in the field. When they have built up their own presence on these sites, you can be sure that they have the experience to do the same for you.

+ **Has nothing but personal social media accounts:** Registering for Facebook and Twitter and using them competently does not make someone a social media expert. The good thing about these tools is that anyone can use them, which is also the bad thing about them. From a professional standpoint, your SMO company should have company profiles on these sites.

+ **Offers a guarantee of results within a specific period of time:** Guarantees are a bad sign in social media because social media is unpredictable in nature. Your SMO consultant can predict possible behavior of the market and suggest a strategy, but hard figures in a certain period of time are outside of the scope of even the most professional of social media managers. A guarantee is a sign of someone who is after your money and doesn't understand the nuances of social media.

+ **Treats social media like a solution for all marketing problems:** Social media is only a small part of marketing online. It is not a replacement for SEO or PPC. Although costs for social media are generally much less than SEO, and definitely much less than PPC, this does not mean that it's a replacement for these tried and true methods. It is not going to save a company that isn't spending money elsewhere in online marketing, and it definitely won't save a company that doesn't have a strong marketing plan in place. SMO should be a small part of an overall marketing plan.

* **Thinks social media is a replacement for offline advertising:** In a similar category as the previous red flag is a social media consultant's focus on making social media a replacement for offline advertising. Offline advertising and media is still valid for the vast majority of markets.

* **No measurable experience in delivering return on investment for social media:** Results from social media take time to develop, but effective SMO has results of some sort. Don't let an inexperienced SMO company tell you that their lack of results is just the nature of social media. Social media can be measured and tracked, albeit in different ways than other forms of marketing, and your choice for SMO should be able to show you ROI experience.

* **Suggests transparency but has no way to implement it:** Transparency is a form of building trust with your audience. Social media networks are places to build your company's reputation and initiate some two way conversations with your market. As Beth Harte suggests in her post "Social Media Transparency: how realistic is it?" transparency may be the wrong term for effective social marketing. A "translucent" approach, where the company representatives initiate conversation without revealing personal details, is the best approach. (Harte) That being said, many social media consultants refer to a professional two way conversation as "transparency". It's a common word in the field, but experienced social media consultants will have a plan for implementing more transparency across the social networks.

* **Suggests that social media be used to achieve total transparency and access to your company:** While this works for some markets, it's not appropriate for all of them. Your market does not need to know what your CEO had for lunch in

order to make a buying decision. Younger markets are attracted to the "peek behind the curtain" of social marketing techniques, but it's not always appealing to all markets. Companies should strive for being "translucent."

+ **Focuses on the tools of social media rather than the underlying concepts:** Inexperienced social media companies will fill their proposal with references to the tools of social media (Twitter, Facebook, Digg, etc) and encourage your company to use those tools. What is lacking in a proposal like this is an underlying marketing strategy that ties these tools together. The tools in and of themselves will not create returns for your social marketing efforts.

It's like expecting your telephone to bring you new leads automatically. Your telephone is a tool that allows you to display your phone number and receive calls for new business. Similarly, Twitter, Facebook, Digg and all of the rest are only tools that can be used by your company, under the guidance of an SMO company. They have no value unless your SMO company gives them value through a strategic marketing plan. At its core, social media sites are tools for communication, just like a telephone. They don't create the conversation, they facilitate it.

+ **Sells the concepts of social media without a workable plan:** An SMO company should be skilled in showing your company what SMO is and how to use it. They should present a complete package for implementation. An inexperienced consultant or company will present some information about SMO basics, refer you to some SMO tool sites, charge a fee and then leave you to figure out the rest.

+ **Pretends to understand all forms of social media:** Ultimately, good social media companies are experienced in advanced marketing principles and can advise your company on how to use new web tools to accomplish marketing goals. They may also specialize in a few forms of social media that they have seen results with and incorporate new techniques as they become apparent. Social media changes too quickly for someone to be an expert in all mediums.

* **Concerned about number of friends, followers, backlinks, etc:** These measurements of "success" of often used by social media newbies to determine how effective social media tactics are. However, the reach of your social media campaigns cannot be determined by pure numbers. For example, there are plenty of people who have thousands of followers on Twitter and don't provide any real value, or make any money from their social marketing. It's much better to have a smaller number of customers that you actually communicate with than a bloated list of people who are unfamiliar with your company.

* **Focused on creating content instead of creating communication:** Social media is conversation and not content. Content creation can be used with social media techniques but it shouldn't be the end goal. For example, your blog uses content to connect with others but your blog is useless from a social media standpoint if you don't interact with the comments or create content in a way that invites conversation. Fifty blog posts on how great your company is don't invite conversation (and it may invite criticism). Quality content is important but it has to be the right kind of content for social networks.

GENERAL QUESTIONS TO ASK ABOUT SOCIAL MEDIA

These questions can be asked of your potential social media company to determine their level of expertise and their approach to social media.

✓ **How long have you been consulting on social media?** Look for a minimum of two years experience with social media. Any less than that and the company may not be able to efficiently manage your social media strategy.

✓ **What is your definition of social media?** The answer should focus on the communications aspect of social media and not the tools themselves. A good answer mentions conversations taking place between people and companies.

✓ **What is your definition of a social media campaign?** Social media campaigns are the process of creating interactive content through the use of social media tools. It gives people a reason to spread the message to others who will be interested in the same content. Their answer should not include references to specific tools or any process that will build links fast (like commenting on thousands of blogs).

✓ **What's the difference between have a social media presence and a social media strategy?** A social media presence is simply having profiles on many of the several social media sites. That can be accomplished in a matter of just a few hours. However, social media strategy is one overall plan for how to use those tools in order to connect with customers and build buzz.

✓ **Do you have any in house published tools or products related to social media?** True leaders in the field have gone beyond pure consultancy to innovation in the field of social media. If they've been able to develop tools and products that other people use, it's a safe bet that they can handle your social media needs.

✓ **Do you have experience using social media with paid search or SEO?** A company that has a breadth of knowledge in the various aspects of driving traffic to your website and increasing your online presence may be able to give you a customized approach that includes more than one approach.

✓ **How does social media relate to SEO?** The answer should refer to social media building relationships that will lead to links in the future. If they deny a relationship between the two, they are too new to social media to be able to help you.

✓ **What is your opinion of quality over quantity?** The social media company should explain that it's important to have quality list members, followers, content, etc., rather than focusing on high numbers. They should explain that social media is a two way street and a good list of followers is better than a massive list that ignores you.

✓ **How do you watch the impact of social media for my company?** Using Google Alerts, TweetReach and Social Mention are an excellent start, but the third party program Radian6 is the gold standard for brand monitoring. Other paid options include Scoutlabs and Trackur.

✓ **Do you have a blog?** Blogging should be the least of your SMO company's experience in social media. Ask for their blog address and check out their posts and their history. Look for a long track record (two to three years of posts), active comments section and frequent posting (at least three times per week).

✓ **What will be your first step for our company?** The first step in all social media campaigns should be listening to what the market is saying. The SMO company in question should talk about listening and evaluating what the consumers are saying about the market and about your company. It's only after they complete these steps that they can develop a strategy to help your company.

✓ **Are you going to create content for my company and represent me on social networks online?** This is a trick question. Social media content should come from your company and not from the social media team. The SMO company should be a coach and facilitator when it comes to using social media tools and implementing the strategy. Ghost written social media content (where a social media company creates content in your name) does not work most of the time and can be embarrassing if your company is "outed". Your SMO company should act as a guide to using these social networking platforms and help your employees use them, not stand in for them.

✓ **What are your customer success stories?** An experienced SMO company should be able to give you success stories and case studies from their previous clients that demonstrate their methodology and their success rate.

THE COMPONENTS OF SOCIAL MEDIA OPTIMIZATION

Ideally, your experience with a social media company will be entirely customized to your unique situation, from the tools that your company uses to the way that they are implemented. There are, however, several components of social media optimization that a legitimate SMO company will use to create their strategy for your company and implement techniques. The following areas should be covered by your SMO company.

CONSULTATION

The consultation phase is the first step of any SMO practice and it's an ongoing practice that needs to take place during the entire time that an SMO company is working with you. A consultation will begin with a needs analysis evaluation of your company and a goal setting session. During this process, your SMO company will ask you about your current forays into the world of social media and what kind of results you want to see from their consultation.

The consultation will help your company set goals for social media usage. There are many different goals that you can reach with social media. In addition to generating leads, increasing sales and growing your customer base, you can also use social media to educate your customers, resource new markets that may be applicable to your product or services and establish yourself in your industry.

Once your SMO company takes note of your goals and your existing social media experience, they will start to listen to the market. They will pay attention to how your market uses social media. Are they avid Twitter users? Do they participate on Facebook? Are there dedicated forums where they meet? This information will help them create the best social media strategy for your company and your market. Since social media is most effective when you participate in conversations that your market is already having, it makes sense to join the existing conversations rather than try to create your own.

They will also look for how your company and your market are being represented online. Using a variety of tools, including Google Alerts, TweetReach and Social Mention, your SMO company will search for the current buzz surrounding your company and the industry. Are they aware of your company at all? Does the market need more information? Are they frustrated with poor customer service from other companies? Are they actively participating in social media events with other companies? Listening to the market will help the SMO company understand the conversation and help your company determine how it can best contribute.

Your SMO company will advise you on the issue of transparency in your social media efforts. As discussed previously, transparency is really being "translucent" (friendly but professional), but there is still a lot of leeway within that parameter. Being translucent should not be a tactic, but a practice for your company. The point of social media is not to advertise your business but to participate in the conversation.

Participation and translucent behavior requires developing a personality for your brand. If one employee is using the social media tools for the company, this will be easier to accomplish than if several people are participating in the strategy under the same banner. Either way, the social media company should help you develop a social media policy. A social media policy will establish ground rules for your company and its dealings in social media. The social media policy will also make sure that social media efforts are in line with the marketing and PR projects of the company.

Social media policies need to address how individuals are using social media, either for the company or on their own personal accounts. Many companies, especially large ones, already have policies for using the phone or email. It's only natural to extend that policy to social media platforms as well. From a legal standpoint, it's important to let your employees know that you have the right to monitor employee use of social media, whether they are using it at home or at work. Social media policy can also remind employees of important anti-harassment, ethics and company loyalty policies that are now extended to social media (i.e.: employees cannot "bash" their bosses or working conditions online without consequence).

In addition to covering employee use of social media as themselves, the social media policy should also cover how the company will be represented through social media. It can be a statement of as little as a page or a detailed manual. If several people in your company will be using social media, it can be helpful to have a training meeting, led with the assistance of your SMO company, which instructs employees on policies and expectations.

Your SMO company should help you develop a policy that explains the purpose of social media and what your staff can do with it to help the company. It should emphasize that they have the right to express themselves, but within reason. They need to be responsible for what they write, particularly on accounts that are associated with the company.

Consultation can also include social media training through workshops, presentations, webinars and online learning. Considering that your company will be implementing the social media methods, you should give preference to an SMO company that includes this type of consultation. It will help your company create effective SMO results more quickly. Certification programs or at least manual for training will put you head and shoulders above the competition.

The last part of the consultation will be a development of how the SMO company will measure your results in social media. Your results will depend entirely upon the goals that you set with the SMO company. After you establish your goals, the SMO company will present you with a method for determining your results.

Keep in mind that because social media is a new form of communication online, the metrics for tracking its effectiveness are new. In the advertising model, money spent is measured directly against the new sales that are created from that advertising or the amount of page views that are created from the advertising. Social media can be measured in the number of users, but more importantly, it should be measured in the amount of direct connections and interactions. In the results analysis section, we'll review a few of the methods that your SMO company can use to track your results and give you feed back on your social media efforts.

RED FLAGS FOR CONSULTATION

+ **Lack of consultation:** If your SMO company tries to apply a "one size fits all" approach to your social media usage, walk away. The social media plan should be tailored to fit your market, your experience and your desired goals. Asking about your budget does not qualify as a social media consultation. At the absolute minimum they should ask you about your goals.

+ **Not using listening as a first step:** After the initial consultation, the social media company should communicate to you that they will begin monitoring your existing social media reach and determining your market's social media activity. If an SMO company immediately advises you to sign up for Twitter, start a Facebook Fan Page or other activity before listening to the market, they don't have your best interests in mind.

+ **No advice on a social media policy:** Your company's ability to implement a social media policy will be key in how your social media efforts will work. A professional SMO company will have suggestions for a social media policy, and may even have previously created examples that you can use.

+ **Lack of basic SMO training:** Your SMO company should train your staff on the basics of social media and how to best implement it. Without this, the strategy that they help you create will be ineffective.

QUESTIONS TO ASK ABOUT CONSULTATION

✓ **What should be my major goals for social media?** This question will help you determine their understanding of social media and how it can be used. Answers should focus on improving your brand image, communicating with customers and getting new future leads. A wrong answer would be "increase your sales immediately."

✓ **What tools will your company be using to listen to the marketplace?** An appropriate list can include paid programs, such as Radian6, but it can also include free sites like Google Alerts, Google Blogsearch, Technorati, Summize and SEO Pro's LinkChecker. A variety of tools is important because it will give your social media company a lot of data to work with.

✓ **Do you offer social media training? What form of training do you offer? Can you provide me with a sample of your curriculum?** A reputable SMO company will offer social media training in some form. Options will include workshops, walkthroughs, webinars or forms of online training. Ask to see a sample of the training to determine if it's of value to your staff, or just a rehash of basic information that you already know.

✓ **What would you include in our social media policy?** Not only does this question test their decision of whether or not the company is planning on using a social media policy, but it will help you determine their knowledge of what a social media policy should contain. Look for a plan to include policies for employee usage of personal accounts as well as how the staff who involved with company social media should represent the company online.

✓ **Will you provide ongoing consultation as our company's use of social media progresses?** It's important that the initial goal setting consultation not be the only consultation that your SMO company provides. Establishing the strategy, looking over results and changing the strategy based on those results are all part of the process.

TIMELINESS

Timeliness is an important quality of social media because if you're using outdated methods, you won't be at the center of the conversation. A good SMO company will have a firm grasp on the nature of the current social media landscape and be your tour guide into this world. There are two major ways that an SMO company can fail when it comes to timeliness: using outdated methods and being too quick to adopt new methods.

Since social media is rapidly changing, it's important that your SMO company stays up to date with the latest tools available to your company. They should be aware of cutting edge sites and help introduce you to new ways of interacting with these sites. They should also be aware of the latest social media trends and keep up to date with the changing field.

At the same time, they shouldn't be pushing brand new sites or practices on your company without having the results to justify their use. Hundreds of new social media platforms launch each year, and not all succeed. Your company shouldn't pour time and energy into participating in a site (or creating a site of your own) that will prove to be worthless. Early adopters of new technology sometimes hit the jackpot, but others end up burned. When it comes to your company's social media profile you shouldn't take too much risk.

RED FLAGS FOR TIMELINESS

+ **Pushing a brand new social media platform on your company:** If a social media company is pushing the "latest thing" on your company then they are trying to use your social media experience as a test for the new technology or website. They should be using their own accounts on these new sites first in order to determine whether it's a worthwhile pursuit for your company.

+ **Using outdated methods or relying too heavily on outdated method:** You can familiarize yourself with what is reliable and useful in social media by browsing through a few key sites. Mashable.com, ChrisBrogan.com and SearchEngineWatch. com are good places to start. You don't have to become an SMO expert, that's the company's job, but you should get to know the top tools so you can understand whether or not the SMO company has your best interests in mind.

QUESTIONS TO ASK ABOUT TIMELINESS

✓ **How does your company stay on top of new social media trends?** Look for a response that includes one of the resources mentioned above, as well as other industry websites that they use to keep their finger on the pulse of new trends.

✓ **What are the most reliable forms of social media that you recommend for your clients?** This will give you a view of what they consider to be reliable and also insight into what you expect their first steps to be with your company, after the listening process. Blogging, Facebook, Twitter, FlickR, SlideShare, YouTube, Linked in and Posterous are all reliable social media platforms that have proven to be essential in the development of a social media strategy.

✓ **What do you do to incorporate new forms of social media into my company's plan?** They should answer that new social media sites and tools are only brought to their clients after testing by their company. Watch out for any sign that they may use your company as a "test company" or that they are eager push new technology on your company without proper testing.

REACH

A social media company's reach is the number of social networks that they are proficient at using. Although your company may be starting small with social media, you will eventually want to expand your efforts. It's important to choose a social media company that is able to grow with you and lead you into other reliable areas of social media.

Your social media company should be experienced at the existing key platforms of social media, as well as be able to create social media spaces that are exclusive for your company if that is within your social media plan. There are many social media sites out there that can be useful for your company, if your market is already participating in them. Having guidance from an SMO company in how to use these sites can be an invaluable asset.

There are many forms of social media you should ask about in your initial dealings with the company. Ask about their experience with any of the following:

o   Blog creation and management

o   Crowdsourcing

o   Facebook Groups, Pages and Apps

o   Forums

o   Photosharing sites (Flickr, Tumblr)

o   Podcasting

o   Ratings and customer service sites (Yelp, Epinions)

o   Social bookmarking and news sharing sites (Digg, Reddit, Delicious)

o   Twitter and other microblogging sites (Tumblr, Posterous)

o   Wikis

However, your SMO company can also assist you with the creation of new forms of social media that are specific to your company. Having a company that is experienced in creating new forms of interaction between the company and the customers is

For example, companies can create interactive applications ("apps") for Facebook that both market the company and provide the user an interesting experience. One such example of a custom Facebook app was Burger King's "Whopper Sacrifice." The app gave users a coupon for a free Whopper if they removed 10 of their Facebook friends. Although the app ultimately closed down because it violated certain terms of Facebook's privacy policy, it was very popular during its short life. (Milian) This example points out why it's important to have an experienced team working on your social media campaigns and tools.

RED FLAGS FOR REACH

+   **Focusing on one tool or one method to implement a social media strategy:** Although your company may start with just one social media tool, the overall social media strategy should not be limited to one tool. An experienced SMO company will have long range plans for your social media efforts which should include the addition of, or creation of, tools in the future.

+   **Unfamiliarity with creating unique social media tools:** If your SMO company's experience is only limited to existing tools, their reach is significantly limited. You need an SMO company that can guide you through both types of social media interaction.

QUESTIONS TO ASK ABOUT REACH

✓ **What are your experiences with these platforms (listed on the previous page)?** This will help you evaluate their experience and ability to grow with your company as you reach into new forms of social media. Look for a company that has strong experience across a breadth of different sites.

✓ **What is your company's design capabilities with regards to social media?** Designing your own tools can be a key part to making the most of social marketing for your company. Your SMO company should have design capabilities and be able to show you examples of apps, widgets, wikis and collaborative events that they helped create for their previous clients.

✓ **Does anyone on your staff handle community management?** If creating a forum or wiki is part of your goals for social media, you'll need a capable person in the SMO company who can easily manage the community and help it grow.

✓ **How much of your staff is responsible for management/consulting vs. building and creating social media platforms?** An SMO company should be balanced in both areas, or at least have trusted resources that they recommend for building social media platforms.

BRAND MANAGEMENT

Brand management is an important aspect of your company's social media efforts. Social media platforms can help you overcome a brand management crisis and respond to the market through new channels. It can also help you establish your brand across the Internet. YouTube and Twitter are both emerging as secondary search engines for many Internet users and your company's silence on these platforms can be deafening (especially if your competition is using these sites).

Why is brand management important? Citizen journalists, dissatisfied customers, organized groups of online protestors or just plain pranksters can do more to hurt a company's reputation in the span of a few days than pre-Internet consumers were able to. Consider the following:

o   The Whole Foods Boycott Facebook group grew to 30,000+ members after the CEO made statements in the Wall Street Journal that were negative toward public health care. (Van Grove)

o   When musician Dave Carroll saw a United Airlines baggage handler mishandle his $3500 guitar, which he found broken after the flight, and the company refused his attempts to have compensation, he released "United Breaks Guitars" on YouTube. The video, which depicted Carroll singing an original song about his experience dealing with United's customer service over nine months, quickly reached over 10 million views and proved to be a public relations nightmare for the company. (Snyder)

o   Domino's experienced a drop in brand value after two employees posted videos of themselves bathing in the restaurant kitchen. The video was viewed over a million times and resulted in a damaged reputation for the pizza chain. (Clifford)

These examples point to the importance of establishing your brand online and then using your online presence to successfully handle a brand crisis.

When your SMO company begins working with you, they should not only be concerned about getting your presence on these popular sites but also how to manage your brand on these sites. The social media policy described in a previous section does much to help your company's social media users establish a solid front across the Internet. However, there are many other steps that a SMO company can help you take in order to establish your brand online.

All of your social media outlets should have the same voice. They can be modified and maintained by different people, but it's important that the same underlying message. A social media editor should be designated at your office in order for the message to be consistent.

You should also use a variety of touch points in your social media use that will help you establish your brand. The name of your company should be used in the same way across all platforms. For example, your Twitter, Facebook and YouTube accounts should all be labeled "Jones Products Inc", rather than "Jones Products" on one and "Jones Products Incorporated" on another. You can also include a tagline or your location as part of your brand identity, but you should make it consistent across the platforms. Your SMO company should also advise you to use your logo wherever possible in order to establish brand identity. The presence of your logo as an avatar on social networking sites will help your company stake its claim online.

Once your brand has been established online through these methods, your SMO company should help you protect your brand. They will monitor your brand and help you learn to do the same. They will help you build a network of strong followers and put you in touch with key influencers in your demographic who will be able to spread the word about your brand. We'll discuss more about these important influencers in the future sections.

When it comes to managing a brand crisis through social media, your SMO company should have a clear plan and experience. Establishing a social media presence well in advance is essential to using brand management wisely. When a brand crisis occurs, your SMO company should target your response to key social media platforms that will get your message the most exposure to your target market. For example, a Facebook page can be used to host pictures, a video or an official statement about the crisis. YouTube is an appropriate place to post videos and can also give your response a high ranking in the search engine results. Twitter allows your company to communicate with the masses in real time.

Since a crisis can't be predicted, your SMO company should help ﹐you formulate a workable plan ahead of time that you can implement immediately. This plan should include checking the complaint for factual accuracies so your company knows exactly how to respond. The response to the crisis should be in line with your company's existing brand and approach to the market, and it should be delivered in a timely manner.

However, a lot of brand crisis issues can be mitigated by being present in the first place. The best approach to stopping a crisis is to invite your customers to engage with you on social media in the first place. Welcoming complaints and criticisms and handling them is one aspect that sets social media apart from other forms of advertising. (Rhodes)

RED FLAGS FOR BRAND MANAGEMENT

+ **No emphasis on establishing the brand across several networks:** Even if your company is not ready to post and distribute videos on YouTube, your SMO company should advise you to register an account there under your company's established brand name. The same step should be taken across all popular social media sites, whether you are currently using them or not, to avoid having a competitor take the name and harm your brand image.

+ **Inexperience with brand management:** Brand management is an essential part of your company's use of social media. If a discussion of brand management is missing in initial conversations with your SMO company, look elsewhere for help.

QUESTIONS TO ASK ABOUT BRAND MANAGEMENT

✓ **What is your brand management monitoring process?** Your SMO company should detail how they monitor your brand (i.e.: using the services of a site like Brandex (http://www. brandindex.com/)). They should consistently monitor your brand's impact through online tools.

✓ **How would you handle a brand management crisis?** The SMO company's plan should include responding promptly across the most important social media platforms, after first verifying the facts. Look for a company that has experience with handling a brand management crisis. Ask for specific examples of how they handled a crisis in the past.

✓ **What services do you offer in the case of a brand management crisis?** In most cases the day to day operations of your social media marketing will be handled by your staff. But when it comes to a brand management crisis, your SMO company should be there to help you overcome the crisis in real time.

Asking this question will help you understand how the company will be there for you.

✓ **How do you determine which consumer statements should be responded to and which should not?** There are a variety of different ways that you brand can be negatively talked about online, however not all of them are worth responding to. Your SMO company should tell you how to determine what deserves your attention and theirs as well. For example, a comment on slow customer service will not pose as much of a threat as a formalized protest that spans across several social networking sites.

METHODOLOGY

Your SMO company should be capable in developing a strategy that will help your company incorporate social media and optimize your social media presence through various channels. There are three steps that an experienced social media company should use to create and implement a social media strategy for your company. (Martelli)

o   Discovery

o   Define

o   Development

o   Delivery

We've already discussed the first step of discovery, which includes goal setting and evaluation of the marketplace. Your SMO company should complete the discovery stage before they suggest implementing any social media steps.

DEFINITION

Next the SMO company will need to define your involvement in the social media landscape. They will take note of the habits of your market through an analysis of the social media sites. They will also review your competitors on social media sites and determine how your company can match their involvement. They will also forecast any potential problems your company may have and develop crisis management plans.

Another important part of the "define" step will be establishing the important messages for your company, the type of content that will be created and what call to action your company should use in your social media efforts. This is essential to maintain cohesion across the various platforms that your company will be using.

The SMO company should define what constitutes a success for your company with social media. Measurement of your brand's impact, the buzz created online from your efforts and your reach are all areas that your SMO company should be able to define and track. You will need to monitor alerts, using a tool like Google Alerts.

Finally, your SMO company should define what social media adoption model has been in place with your company, and what should be used from here on out. There are many adoption models possible, but information technology consultant Dennis McDonald explained the most popular options succinctly in his post "What Social Media Adoption Model Are You Following?"

The four options are:

o   Top down – This model requires the company's leader to be the primary adopters of the technology. For example, a blog from the CEO or having the upper management maintain Twitter accounts. This typically happens when the management sees the value of social media and wants to be the first to establish the corporate identity in social media networks.

o   Bottom up – As an opposite of the first adoption model, this one is used when workers being the first users of social media. They may use free social media sites in order to complete their job, or they may use them on a personal basis. Whatever the reason, it becomes clear that corporate policies are a requirement for social media usage and employees will be key allies in making social media work for the company.

o   Inside out – If your company has been following an inside out adoption strategy, you already have social media tools in place for your company for internal communications (for example a group wiki that helps employees stay on target with policies and projects). Social media tools are familiar territory for your company, but your staff needs to learn how to incorporate them with the public.

o   Outside in – Your company does not use social media however, your competitors are making use of the technology quite publically. Driven by the current marketing culture, your company has decided to investigate how social media can help.

(MacDonald)

These adoption models will make a big difference in how your SMO company will define your participation in social media and assist them in developing a plan for your company to follow.

DEVELOPMENT

In the development stage, the SMO company will help your company develop the plan of attack for your use of social media. They'll select the platform (or platforms) that your company will be using to reach your target audience and how your company will engage in these platforms. This can include:

o   Establishing and updating a corporate blog.

o   Creating and maintaining an official Facebook profile or Facebook fan page.

o   Assigning a specific number of employees to sign up for Twitter and build follower lists.

o   Establishing wikis to share important information for the market.

o   Creating and distributing weekly podcasts.

o   Incorporate social bookmarking to support the identification and tagging of relevant information sources.

o   Creation of widgets or apps for your target market to use.

o   Any variety of social media tools and techniques that have been discussed in this chapter.

If you are a newcomer to using social media, they should outline a plan for the growth of your social media campaign from the initial platform out to additional platforms over the course of several months to years. Your growth in social media should be outlined in their strategy with benchmarks and evaluation points all along the way. Typically, this is covered in a detailed proposal that the SMO company will give you once they have investigated your niche and your current dealings in social media.

It's important that your SMO company give you guidance not only in using the platforms, but how to best use them for your target audience. With your marketing messages in hand, they should create a brand identity that will be represented across all of the platforms. They should create key message points that will be used to deliver content to your contacts on the social media sites. This can include developing a publication schedule for a blog, maintaining a list of reliable resources for gleaning relevant news to share through the platforms and establishing a schedule for social bookmarking practices.

Another essential part of the development process will be identifying and connecting with key influencers in your market. This is a technique that sets apart experienced SMO companies. Influencers are people involved in social media, in your niche, who are able to spread your message and recommend your company. They've taken the time to build a list of followers, friends or readers (depending on the network) that trust their recommendations and listen to what they have to say.

Influencers can help spread your message far and wide across social media networks, with less effort on the part of your company. They act like a megaphone to the niche and get your message out there more quickly. Since the influencers have already built trust with those that they influence, your message delivered through their mouthpiece will have a huge impact.

Contacting and working with key influencers in the niche should be part of your SMO company's plan for your social media efforts. Your company can connect with key influencers via many different social networks and then use those connections to distribute your content and marketing message. Sharing information, posting content that can be reprinted by others, guest posting on an influencer's blog and offering to be interviewed for podcasts are all examples of how companies can be available to influencers and create a ripple effect across social media platforms.

In addition, your SMO company may also suggest developing proprietary social media platforms hosted by your company, or as apps or widgets for other platforms. Developing independent social media channels or tools can help your company increase its presence online.

DELIVERY

The delivery phase includes your company delivering the content that your SMO company has outlined and then seeing the effect it has on your market place. By engaging in the conversation with your market, you'll begin to learn more about how they use social media and how you can best engage with them. No matter how experienced an SMO company is, there will be refinement of your social media practices as you get feedback from the marketplace.

Social media offers your company a unique opportunity to ask the market exactly what they want. Continue to build relationships in the delivery phase through building conversations and increasing your network of contacts.

These four areas – discovery, definition, development and delivery – are all key points in any methodology that is used by a professional SMO company. Look for these characteristics and you can be sure that your social media efforts are in good hands.

RED FLAGS FOR METHODOLOGY

- **Strategy is missing any one of the key areas discussed above:** Anyone could discover your social media presence or help you develop a profile on one site, but only an experienced SMO company will be able to pull them all together and give you a complete plan for social media usage.

- **Strategy does not align with your adoption method:** If your company has not been using social media for corporate use, it doesn't make sense for an SMO company to suggest that employees begin using social media immediately (adopting an inside out). If the CEO does not want to participate in social media, a top down strategy will not work. Make sure your SMO company knows your company's experience with social media and that they present a logical solution based on your experiences.

+ **Lack of contacting influencers:** Influencers are key to making an efficient use of social media methods. Your SMO company should be locating and helping you get in touch with these influencers in your social media channels.

+ **Trying to complete social media tasks for you:** An SMO company acts as a guide to social media, and not a substitute for your company. Although initially there may be some modeling done by the SMO company (i.e.: sample blog posts or Twitter updates), be cautious of a company whose idea of methodology is to take over the reins of your social media participation.

+ **Pushes app or widget development as the only solution for social media:** Your use of social media platforms should extend well beyond "cool tools" that your SMO company can create for you. If your SMO company presents this as a sole solution, they may not be experienced in other methods of social media, meaning that they won't have the capabilities to assist you with promoting the tool effectively across social networks.

QUESTIONS TO ASK ABOUT METHODOLOGY

✓ **Can you share your strategy for social media development?** The answer to this question can vary, but it should closely reflect the four step method shared above. The strategy should not be a list of tools but a detailed plan for establishing your company on social media sites, interacting with the community and posting content. Look for a company that can show you specific examples and case studies of their strategy methodology.

✓ **How do you incorporate my existing social media approach into your strategy?** This question staves off companies who would rather you "wipe the slate clean" with regards to social media. A smart SMO company will work to improve your existing usage of social media, no matter what it is.

✓ **How will you ensure that your social media strategy complies with our legal requirements?** The SMO company should encourage you to develop a social media policy, but most importantly they should do their due diligence before creating and presenting the strategy for your company. The SMO company should ask about any existing policies regarding social media usage or advertising online, particularly to regard with statements that can be made about products or services.

✓ **What is your process for identifying and connecting with influencers in my niche?** The SMO company should ideally be familiar with your niche from previous experience with other companies and be able to identify the key influencers. However, if they are not familiar with your niche, they should describe doing a to find bloggers, speakers, authors and other online leaders who are using social media sites to connect with their audiences. They should be actively participating with their audiences on a consistent basis (i.e.: frequent blog posts, consistent Twitter updates, etc). Your SMO company should suggest getting on their contact lists in whichever social networking platform you're using, and making key contributions to their experience (i.e.: leaving content on blogs or responding to Twitter posts).

✓ **How do you make sure that transparency and authenticity are maintained during influencer outreach?** The SMO company should suggest that the first interactions be purely to the benefit of the influencer (i.e.: support for their site, sending information their way). Just like in real life negotiations your company will have to show through social media how you can benefit the influencer.

RESULTS ANALYSIS

Social media, like all other business practices, should produce results in the bottom line in order to justify keeping it as part of your company's advertising and marketing efforts. However, the ROI on social media is not as cut and dry as it is with other models. There is not a direct correlation between social media usage and sales as one would expect. Only by using social media metrics and looking at the overall impact of social media on all other aspects of the business can your company see the impact of SMO.

Before you look for an instant impact to your business, understand that social media needs a long term approach. Since it's primarily about building community and becoming involved with your niche, it is a process that takes time. However, this does not mean that it's above measurement and analysis.

Collecting friends on Facebook, getting traffic to a blog, increasing Twitter followers and connecting with influencers in your niche is terrific but these actions aren't directly creating profits. What they are doing is creating opportunities for increasing your revenue. Your SMO company will be able to use various metrics to show the impact of your social media usage. The following categories of activities will help your SMO company track the real ROI for your efforts. (Blanchard)

GOALS AND OBJECTIVES

Setting goals and objectives for your social media involvement is important to analyzing the results. We've discussed that part of the process in depth in earlier sections of this chapter. Once your social media strategy has been running for a while, your SMO company should be able to assist you with determining if those goals are being met. Goals are constantly being revised and expanded to respond to the feedback received from the community.

ESTABLISHING BASELINES

Before your SMO company begins rolling out the strategy for your company, they will take a baseline reading of various metrics for your website. These will include sales, traffic, new customers and various other metrics to be determined by your SMO company based on their experience. This will be the starting point for your social media involvement to determine what is working and how.

CREATE ACTIVITY TIMELINES

Your social media involvement should be tracked on a timeline that will help your SMO company correlate your social media efforts with increases in sales. For example, keeping track of what steps are taken by your team on a weekly basis will help you and your SMO company determine which actions are having the most impact. Usage of new platforms, press releases, podcasts, distributing new content, comments on important blogs and conferences should be tracked on this timeline for future usage.

MEASURING RESULTS AGAINST THE BASELINE AND SOCIAL MEDIA PRECURSORS

After letting the social media usage run for several months, your SMO company should begin to measure your results against your social media usage. They will compare your timeline and their own social media metrics tools against your sales revenue, new customers, number of new transactions and the amount spent per transaction. If there's an increase in these results, they'll look back to what happened on the social media side of things to pinpoint the reason for the increase.

By layering the various metrics, your SMO company will begin to see a pattern emerge. They'll analyze:

o   Social media activities

o   Social data (number of comments, friends, followers, etc)

o   Brand popularity

o   Website data (traffic, backlinks, etc)

o   Sales transactions

From these metrics, a picture will begin to form about what practices work best for your market and your company. This will inform your future strategy and help your company determine where your time is best spent.

RED FLAGS FOR RESULTS ANALYSIS

+   **Claiming that social media has "no true ROI":** This is a commonly held misconception because the true ROI for social media is not seen in a one-to-one ratio like other forms of advertising and marketing. For example, your company can expect that for every 5,000 leads sent to your website, you'll have ten sales. However, 5,000 friends on Facebook does not lead to a specific number of sales. Through the process outlined above, an SMO company will be able to tell you the ROI for your time and money used in social media.

- ◆ **Lack of professional metric tools:** Your SMO company can tell a lot about your social media impact with free tools, but you can use them yourself. You're paying an SMO company for their expertise and this should include paid tools to fully analyze your social media results.

QUESTIONS TO ASK ABOUT RESULTS ANALYSIS

- ✓ **How do you measure ROI?** The SMO company should explain that it's a detailed process, but be able to give you a basic overview (like the one described above). If they are resistant to answer the question because they feel social media is a goal in and of itself, they don't have experience running professional social networking campaigns.

- ✓ **Have you developed any proprietary metrics that you use to enhance your analysis of social media impact?** Many professional social media companies have developed ways to analyze data that are specific to that company. If they have, ask for case studies and examples of these metrics in use.

CONCLUSION

Social media optimization is a rapidly changing field and some aspects of it are very experimental. It's important to find a SMO company that exhibits proficiency in the basic elements that have been established. However, it's also important to understand that even those at the top of their game are adapting to the new methods, tools and practices in social media optimization. Looking over the points in this chapter will help you identify those providers who are on the forefront of this exciting field.

CHAPTER SIX:

LANDING PAGE OPTIMIZATION SERVICES

Landing page optimization is an essential part of having a successful website. The landing page is the first page that your visitors arrive at once they click on a link. It's their introduction to your site, your company or your product. Optimizing the landing page with the help of a trained professional can increase your conversion rates, and therefore, your sales.

An experienced landing page optimization (LPO) company can make your pages more effective. They will analyze your current landing page and use your site's past performance to determine what needs to be changed. The LPO company will develop a plan of action to adjust the design, content and layout of the landing page so it is more effective. This may include a total redesign and subsequent testing, or split testing several small elements that create a better conversion rate.

Landing pages are essential to the success of your website. It doesn't matter how much traffic you are driving at your website if that traffic isn't converting. Changing the conversion rate by just 1% will increase your results and make your advertising dollars more effective. However, landing page optimization can be a complex process. It involves understanding your target market's motivations and translating those motivations into technical data that will produce the desired results.

The following chapter will expose you to the basic concepts of effective landing page optimization, so you know what to look for in the services of an

LPO company. We'll discuss the correct methods, strategy and execution for LPO that will produce results.

LANDING PAGE OPTIMIZATION BASICS

Landing pages are the first pages that your visitors encounter after they've taken an action (like clicking a link or completing a form). Visitors are delivered to your website in one of several ways.

o   Clicking on pay per click ads that are displayed on search engine results pages or the content network.

o   Clicking on banner ads that are displayed on another website.

o   Clicking on a link in an email marketing message.

o   Clicking on a link in the sidebar of a blog or a blog post

o   Clicking on a search engine result

(Clark)

There are three basic types of landing pages that you will use in your online marketing activities – the main site, microsite and stand alone site. (Ash) The "main site" is part of your main company's website. The main site landing page will be within your company's site, but not necessarily the home page. A good example is a subpage of your main website that is designed to educate your visitors on the importance of proper patio drainage and motivate them to click on the link to learn more about your product.

A microsite is designed to market to a specific group of people that your main site does not reach. For example, a company that sells information on taking better photographs may create a microsite to market their material to parents who want to take photographs of their children. This market is not reached on their main site, which is directed toward amateur freelance photographers. The microsite gives them the opportunity to fine tune their message to speak directly to that market, which may not relate to their main site. A microsite is a smaller site that has just a few pages designed for a specific purpose. It can be branded as part of your main site or it can remain separate.

Finally, there is the stand-alone page which has information that is related to a specific offer. There is one page, one call to action and one purpose. For example, a sign up form for a one time webinar, a special sale for a specific product or the release of an industry report can all benefit from a stand-alone page.

The type of landing page you use will be informed by the decision you are asking the visitor to make and the market you are targeting. Your landing page is supposed to be guiding your visitor toward a specific action. It could be to make a purchase, read an article, sign up for free information, click on to the next page, etc.

One of the primary functions of landing page optimization is to define and refine the purpose of your landing page. If you are requiring your visitors to do too much when they arrive on your site (opt in to your list, buy a product, read information, etc) they get confused and overwhelmed.

The attention span of a website visitor is very short, and you have to be crystal clear about the purpose of your landing page when they arrive. A landing page that is designed to get visitors to opt in to receive a free white paper will be structured very differently from a page that is designed to sell an energy drink. Determining the purpose of the landing page is essential to creating the design and the content.

Some examples of the purpose of your landing page can include:

o   Selling a physical product

o   Selling virtual products

o   Educating the visitor to make a decision

o   Generating leads

Another key factor in the development of a landing page is determining who the audience is and how to motivate them to take the action. Successful landing pages are created with a specific audience in mind.

Take the energy drink landing page for example. A landing page that is the result of a PPC ad for "sports nutrition" will be different from one that is the result of a PPC ad for "low energy". The markets (athletes and people with low energy) have different needs, different psychological push buttons and different requirements for what they are looking for in an energy drink. A manufacturer who is targeting their drink toward athletes will have a much different landing page than a different manufacturer who has created an energy drink for the average person who needs more energy.

Market analysis is essential to creating an effective landing page. If you are not familiar with your market and their motivations, it will be difficult to create processes on your website that will lead to conversions.

After establishing the purpose of the website and the target market, an LPO company will help tweak elements of the page in order to increase conversions. This is referred to User Centered Design (UCD). UCD is a process that was initially identified by Dr. Don Norman, a leading usability expert. The purpose of UCD is to create a website that speaks to the average user that will be arriving on that site. It narrows down the steps on the page to create a high level of conversions for each specific task.

The following sections will go over the steps that a LPO company should use in order to create an effective landing page and increase your conversions. In addition to these criteria, you can review the latest listings for landing page optimization companies from topseos.com at http://www. topseos.com/rankings-of-best-landing-page-optimization-companies.

GENERAL LPO RED FLAGS

+ **Familiarity with SEO and not LPO:** Search engine optimization and landing page optimization are closely linked, but they are not the same thing. SEO is tweaking the code, copy and layout of a website in order to increase search engine ranking results. LPO has to do with the user's actions on that page. Your LPO company should be experienced in LPO specifically, and not just SEO. Inexperienced companies may try to establish themselves as experts in LPO when in actuality they just have SEO experience.

+ **Downplaying the importance of LPO:** Multi-service Internet marketing firms may offer LPO as one aspect of their service, but it should still be an important part of their plan to make your website effective. Watch out for a company that emphasizes traffic building techniques over the optimization of your site. Remember, increased traffic doesn't help your site if that traffic is not converting.

+ **No help in defining your type of page or ideal customer:** An experienced LPO firm should be well versed in the basics of what it takes to make a page effective, including defining the type of page, the decision to be made on that page and the ideal customer.

+ **No experience working with sites in your field:** LPO is closely linked to market research and the motivations of your ideal buyer. Without sufficient experience in a related field, your LPO company will be less likely to nail down the specifics of what it takes to get that group to make a decision on your page. Find and hire a company that already knows what makes your customers tick.

GENERAL QUESTIONS TO ASK ABOUT LPO

✓ **What increase in conversions can I expect from your services?** Any credible company will let you know that conversion rates are particular to your site and your individual circumstances. They will not give you a specific or guaranteed amount of increase for their work with you. Alternatively, they will explain that their services will help you set goals for conversion and attempt to reach those goals through optimization techniques.

✓ **How quickly will you be able to meet the conversion goals that we set?** This is another trick question that will weed out unscrupulous companies. Any company that claims it can meet your conversion goals within a month or two is just trying to take your money. A quality LPO company will explain that their services will extend for months past the initial implementation of the strategies in order to track results and make further refinements.

✓ **Are you familiar with user centered design?** This question will test their knowledge of user centered design concepts. Any experienced LPO company should be able to answer this question succinctly and knowledgably.

✓ **What methods will you use to increase my conversions?** The LPO company or consultant should lay out a plan that involves collecting current conversion data, defining the ideal customer, discovering their motivations, using design and copy in order to speak to those motivations, making changes and tracking those changes for effectiveness. All of these steps are necessary for effective LPO. Look for a specific methodology and not just testing and tracking.

✓ **What are the risks involved with landing page optimization?** A company should explain that there are some risks involved, including lowered conversion rates for a short time. There is always a slight chance of this when anything is changed on the landing page. Without being aware of it, you may already have several elements on the page that are optimized to their fullest. Your LPO company won't know this until they begin changing those elements and noting the results. In addition, the optimization process will most likely include changing the site design and the copy. This may result in changes in your website's standing in search engine results.

✓ **Do you have experience working with my industry?** This is important to the LPO company's plan for increasing your conversions. Familiarity with the industry will help them identify and nail down the ideal customer and their motivations in order to inform their conversion choices.

✓ **Will you work with my shopping cart program?** This is important if you have an ecommerce site. Your LPO company should have familiarity with your shopping cart program so they can use it to effectively increase your conversions. If they haven't worked with the specific program that you use, they should at least be familiar with working within the constraints of a shopping cart template.

✓ **What testing software do you use?** Testing is an essential component to optimization. Google Website Optimizer is a popular tool, but there are also other programs like SiteSpect, Verster, Hiconversion, OnDialog and SplitAnalyzer that are also used by professional landing page optimization companies.

(Allem)

SITE ANALYSIS

Site analysis allows your LPO company to establish baseline metrics for your conversion rate. It also helps them identify key areas where your visitors are leaving your site and opting out of the decision making process. It is essential that your LPO company use site analysis as an initial part of their work with you. Before they can start implementing steps for improvement, they have to know where you stand.

ANALYZING THE MARKET

As noted in the last section, your market and your landing page are closely linked. Your LPO company should help you define your target market and adjust your landing page to meet that market. You'll let them know about your market and they'll analyze your landing page through the lens of that market. How well does it meet the needs of that market? Does it speak to their motivations?

You should provide demographic data that your company has developed for your ideal audience. In addition, the LPO company will glean demographic data from your website's traffic data (see the next section). Your LPO company will look at where your traffic is coming from. Is it through PPC ads? If so, what keywords are garnering the most click throughs? Are a series of blogs sending traffic your way? If so, who are the readers of the blog?

A professional LPO company should look beyond the demographic data and begin to develop a user profile based on the psychological needs of your ideal users. Many companies have websites that are not appealing to users, or they are appealing to the wrong type of user. That's why it's so essential for your LPO company to do market analysis testing to understand your ideal customer, as suggested in the previous section. The page should be viewed through the lens of the ideal customer and created to meet their needs.

ANALYZING THE DATA

If your website has been using a tool like Google Analytics to track your results, you'll need to give this data to the LPO company. If a tool like this is not in place, one of the first steps your LPO company should take will be to establish tracking tools that give detailed data on how the visitors are spending time on your site. Even without this detailed data, they should be able to glean quite a bit of information from your website stats, your sales figures and your opt-in rates. However, in depth analytics software allows an LPO company to get more done with your site. If you haven't yet hired an LPO company, consider installing Google Analytics or another similar software program in order to get some data for them to mine through.

For example, with web analytics software your LPO company will be able to determine if the traffic is primarily coming from new visitors or returning visitors. This is an important distinction because returning visitors have already been exposed to your marketing message. The actions of the returning visitors will help the LPO company determine if your message is strengthened upon their second visit (they make a purchase or take an action) or if it is weakened.

The main pieces of information they'll be searching for are your conversion rate, your cost per conversion and the bounce rate. Conversion rate is the amount of traffic versus the amount of conversions. For example, a website receives 929 visitors in a single day, which produces two sales of $129 each. The conversion rate is .2% (2/929). In the same day, the company spent $35 in PPC ads. The cost per conversion is $17.50 ($35/2).

The bounce rate can only be determined by an analytics program. This figure measures the amount of website visitors that are arriving on the site and leaving after visiting just the first page. It indicates the visit quality. (Google) Do your visitors hang around your site long enough to explore it, or do they click away? These are questions that your LPO company will be using the data to answer. They should have a goal to keep your bounce rate under 50%. Anything over that means that they need to better define and enhance your landing page.

The goal is to have your website visitors interact with your website as much as possible. The length of the visit will increase as your site is developed to meet the needs of your market. If you find that a specific group of users are spending large amounts of time on your site, you can make them a "break away" group and develop content to specifically meet their needs. (Ash)

ANALYZING THE PAGE

In addition to analyzing to existing traffic and conversion stats, they should take a look at your website design and copy to determine what needs to be changed. Landing pages should follow common sense guidelines for user interaction, but unfortunately common sense is not always so common.

Generally speaking, most users online are impatient. They will scan through your landing page looking for the information they need. They will look at certain types of images and will look for an opportunity to click on a link or button, or they will click away from your site. The types of images, text and "click opportunities" will be determined by the type of user that is visiting your site.

In addition, your site should be easy to understand and navigate. The information architecture (how the information on the site is organized) should be visible and support what the user is there to do. Your landing page should be consistent with the rest of your site but also targeted toward the referring traffic.

There are several common mistakes that web designers, copywriters and company owners make with landing page design. In fact, Tim Ash, author of "Landing Page Optimization" calls them "The Seven Deadly Sins of Landing Page Optimization."

They are -

o   An unclear call to action - Your landing page needs to give your audience direction and purpose. Many landing pages miss the mark in this area by relying solely on a graphic splash page for their landing page, or by directing all of their traffic to the homepage. As your LPO company helps you define your ideal customer for this page in the strategy development phase, they'll create an effective call to action.

o   Too many choices - A narrowly focused landing page with one direct purpose will convert better than a "hub style" page that has links to too many different options. If your landing page falls under this category, your LPO company will likely streamline the page's purpose so it only presents one choice for the visitor to make (signing up, purchasing or visiting another page of the website).

o   Asking for too much information - Requesting that your customer fill in their name, email address, street address, phone number and other personal information can be a real detriment to conversions. Your LPO company should advise that you limit your request for information to just a name or email address, unless you've built a lot of trust with your market.

o   Too much text - When a visitor lands on your site, they are looking for specific information. Assaulting them with large blocks of text that they need to read will make them click away. Long, explanatory paragraphs do not belong on a landing page, or on any page for that matter. The attention span of most website visitors is very short and they won't tolerate long blocks of text.

o   Not keeping your promises – This specifically applies to landing pages that you'll be using with pay per click ads. If your pay per click ad promises the best prices on gluten-free bake ware, and the landing page is a subscription site for gluten free recipes, you aren't keeping your promise to the customer. That's why it's essential to target your landing pages to address specific needs.

o   Visual distractions – Just as too much text can distract from the clear purpose of the website, too many graphics and a complicated layout can decrease conversions. Your LPO company should analyze your site's design and make suggestions for streamlining the look of the site for better conversions.

o   Lack of trust – The level of trust that your market has with your website is directly related to how many elements of your site enhance that trust. For example, security bugs, certifications, credit card emblems and other professional designations should be incorporated into the design of the page in the first screen shot.

(Ash)

In the site analysis process, a LPO company should review your website and take note of any of these major problems with the design. For example, with web analytics they'll be able to determine which content is getting the most views, the most backlinks, etc. They will also be able to tell what content on your pages is resulting in the most exits. Determining what type of content appeals to your market and what type of content makes your market leave your website will inform content creation in future sections.

TOOLS FOR ANALYZING

LPO experts should be able to identify major problems with your site from their knowledge of your target market and their knowledge of the principles of LPO. This is why it's helpful to hire an LPO company that already deals with your industry. They have seen hundreds of poorly designed websites in your niche and can use their experience in order to help you create an effective landing page.

In addition to their own expert evaluation, your LPO company should use other people's input in order to determine the successes and failures on your website. Full scale usability testing is the most comprehensive (and most expensive option). When an LPO company conducts usability testing it will recruit users who meet your needs and have them interact with your website. The company will analyze their interaction with the site and look for specific tasks or measurements (i.e.: how long the subject stayed on the first page or how many pages they viewed). It can also have the test subjects tell the testers how they feel about the site and their initial responses to its visual design and copy.

This testing can be done in large focus groups rather than on an individual basis. In a focus group, your LPO company will lead a discussion on their impressions of the website and take note of the general response to the website. Although the results of focus groups can be skewed by outgoing participants, it can be very helpful to have this type of testing done.

Eye tracking studies are also an effective means of analyzing the usability of a website. Eye tracking studies will use live website visitors to determine where the average user will look first on a webpage. An eye tracking study will result in a "heat map" that shows which areas of your website get the most attention. Based on this heat map, your LPO company should be able to streamline the content and design to increase conversions. Even if your LPO company does not conduct an independent eye tracking study of your specific website, they should be able to make some changes based on what eye tracking studies have proven in previously. For example, eye tracking studies have determined that, generally speaking, users follow an "F shaped" pattern when it comes to viewing a webpage's content.

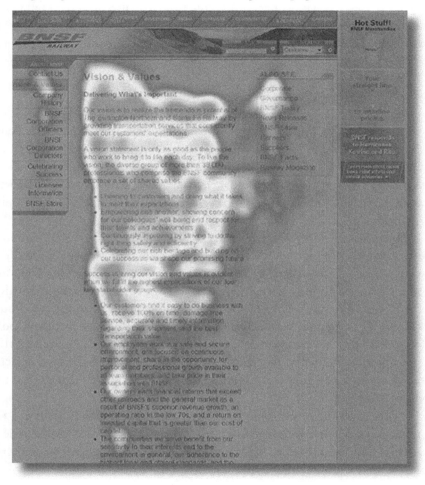

(source: http://www.useit.com/eyetracking/)

The red/orange areas show where the website visitors are spending the most time looking. It indicates where most of the users looked at that part of the page for a fraction of a second or more. The yellow areas also attracted user attention, but less than half of the users spent a significant amount of time there. The blue areas got a glance from a small percentage of users.

The classic F shape pattern can be used on your website to increase the visitor response and improve your conversions. There are heat mapping tools that can analyze your website and determine where visitors are most likely to look and click. Heat maps can help optimize landing pages, minimize the likelihood of your users leaving the shopping cart and make web usability testing easy.

Your LPO company should be familiar with heat map tools. Some of the most reliable tools are CrazyEgg, ClickDensity, phpmyVisites, Feng-GUI and Conversionstats. All of these tools are available for free or a low subscription cost, so in theory you could utilize them yourself to identify key areas that need to be adjusted on your landing page. Your LPO company should use these tools or advanced versions of these tools to determine where the most amount of time is being spent on your website.

Your customer service representatives can be a great source of information on website usability. They hear the most about your website and the problems that users have with the website. If you don't have a customer service department, watch for trends in help desk tickets or customer support emails. This can be a goldmine of information for your LPO company.

Finally, your LPO company may use formal surveys targeting your ideal type of client to determine how the site is doing. They will either install a survey component on your site and let it run for a while or coordinate the hiring of a survey company to get valuable feedback on your landing page from your target market. Be sure that your LPO company is using random members of the target market and not those who have already completed the conversion action. Those who have completed the conversion action can't tell you about what is not working, because obviously the landing page worked the way it was supposed to for them.

RED FLAGS FOR SITE ANALYSIS

* **Jumping into optimization with no site analysis:** A LPO company should not use a blanket approach to your website. Even though they are experienced in increasing landing page usability, they should complete site analysis first in order to determine the best way to use their LPO "bag of tricks." They should also implement web analytics tools immediately if your site does not currently use them.

- **Ignoring the "Seven Deadly Sins":** Inexperienced LPO companies will focus on one aspect of optimization that they have had success with in the past and ignore the rest of the important usability issues. If you find that a LPO company is focused on making sure that your website design is flawless without touching your website copy, they do not have enough experience to help you.

- **Lack of testing with the market:** LPO companies should know a lot about usability, but your website is unique. It's essential that they use one of the methods above (heat maps, surveys, focus groups or customer service feedback) in order to gauge how your market is interacting with your landing page. This will mimic the interaction that your page will have with the real world.

QUESTIONS TO ASK ABOUT SITE ANALYSIS

- ✓ **What data will you use to analyze my landing page's effectiveness?** The LPO company should respond with the name of a specific third party program (which you can later investigate) or Google Analytics. The data they will use from these tools will include traffic, conversion rate, cost of conversion and bounce rate, among several other factors. They should take an initial baseline and then compare that baseline to the same data after they implement their strategies.

- ✓ **What are some initial steps you'd take with my landing page, as you look at it now?** Before delving into the analysis of the site, your LPO company should be able to tell you which areas of your site need immediate attention at first glance. Look for mention of one of the "seven deadly sins" in their analysis.

- ✓ **Will you be using heat map technology will the optimization of my landing page?** With recent developments in heat mapping software, this should be a given with any professional LPO company.

✓ **What supplementary methods of analysis will you use to help optimize my landing page?** Look for a LPO company that will incorporate third party input from focus groups, surveys or market studies. This will result in more effective landing page optimization techniques.

STRATEGY DEVELOPMENT

The analysis step sets up your LPO company for developing a strategy to optimize your landing page. The primary driving force behind strategy development will be increasing the likelihood that the ideal visitor will take the required action. There are two main questions that your LPO company should answer when developing a strategy for your site:

o Who is the typical user?

o What are they there to do?

These two questions will be answered through the development of a persona that represents your company's typical customer or the identification of roles that the typical customer takes. In addition, strategy development should include the discovery of the most likely task for that persona.

PERSONA AND ROLE MARKETING

Persona marketing is an effective technique for increasing your website's usability when you have customized landing pages. Your landing page may receive a wide variety of traffic, but an LPO company should determine what segment is most likely to take action and then develop a plan to target that segment.

A persona is a fictitious dossier that will inform site optimization strategies. It will give your LPO company someone to market to specifically. Most companies would like to reach as many people as possible and might balk at the idea of targeting a landing page to one specific type of user. There are two main reasons why persona marketing is essential for the success of your landing page.

1. Attention spans are short online and that landing page needs to make a big impact in a short period of time – sometimes as

quickly as a few seconds. Your site is much more likely to make an impact if it is created to appeal to your buying audience.

2. Persona marketing helps weed out the users you don't want as much as it appeals to the users that you do want.

Your LPO company should develop a detailed persona that will be used as a template for your company's landing page optimization. Here's an example of what a persona dossier would look like for a home improvement landing page for a video series on building a deck.

Henry Handyman

Henry is a 40+ homeowner who has desire to increase the value of his home by building a deck.

He is experienced with home improvement projects.

He has completed many home improvement projects and is confident he can build a deck with the right guidance.

Familiar with the cost of having a deck built and wants to save money.

Had a bad experience with a contractor in the past.

Married, 2 children.

Middle management, white collar job.

Familiar with using email and reading information online – uses the Internet for work.

Owns major power tools needed for the job.

Quote: "I'll do it myself, as long as you show me how."

As you can see, the persona includes demographic information (like age, location or interests) but it also hints at behavior patterns that define how the persona will interact with the site.

o He is familiar with online technology for work, but purely at a consumer level – meaning the site shouldn't have a lot of overwhelming graphics or ask the user to input a lot of personal information.

o He does not trust contractors and is looking to save money and avoid problems by doing it himself – meaning that the content should be structured to emphasize that the videos are a good deal and that they are "insider secrets."

o He wants to do things himself – which means the site shouldn't talk down to him or make him feel inferior.

o Online video is new to him – he will need help in overcoming the objection that the videos are hard to download or that physical videos are a better investment.

Personas work well for tightly focused niche site landing pages, however, they may not be as effective for larger scale landing pages that may have visitors across many different cultural backgrounds. If your landing page is targeting a broad audience, your LPO company should develop roles to define your audience and help clarify the optimization of your landing page.

Roles for a home improvement tools online store could include:

o Hobby craftsmen who do home improvement projects on the weekend.

o Independent home improvement specialists.

o Small scale home improvement businesses.

o Novice homeowners who need a specific tool to complete a simple job.

All of these roles have different backgrounds and in order to meet their needs, the site can't be too specific. The site can't heavily rely industry specific terms because that would leave out the novice homeowner.

The key difference between personas and roles is that a person's role can change depending on what they are doing, while a persona is usually a fixed identity with a specific agenda, job, age, etc. The novice homeowner may be inexperienced with home improvement projects but considered an expert when picking out a new home computer system.

Analyzing roles and personas will allow your LPO company to develop a strategy that is targeted directly at the needs, tendencies and desires of these groups. Your landing page only has a split second to attract and keep the attention of your visitor. By tailoring it as close to possible to your ideal customer, your LPO company should be able to increase your conversions.

DETERMINING TASKS AND SEGMENT OPTIMIZATION

The personas or roles that are visiting your website are there to do something specific. Depending on the nature of the landing page, there may be opportunities for them to do one of several things. Your LPO company should determine the tasks for the persona or the role and then retrofit your landing page to meet that need.

For example, an accounting services site may have several different role types visiting the site. However, for the landing page in question the SEO, PPC and other traffic sources are all coming from small business professionals who are looking for tax help. They need information and accounting services. The landing page should be targeted to help that specific group accomplish that specific task. The easier your LPO company makes it to achieve the task, the higher your conversion rates will be. However, it is impossible to determine the right task for traffic without implementing "segment optimization."

As indicated in the example above, specific traffic channels should be driving toward specific pages on your site. Think about it this way – you are searching for a Black and Decker coffee maker and click on a link from Amazon.com. Rather than transporting you to the Black and Decker coffee maker in question, the link takes you to the homepage. The homepage isn't related to your needs, so you click away and Amazon lost out on a customer.

Obviously, Amazon.com doesn't do this to its customers, but website owners can learn and important lesson even if they don't have a product website. Your traffic should not be routed to the homepage and your website should include pages that are each individually optimized for specific roles and specific tasks. Segment optimization utilizes landing pages that are designed to meet the needs of a part of your audience. (Brinker)

The accounting site from the previous example would develop separate landing pages for each of the major user types. These user types would be determined from their traffic, their market research and their LPO company's expert advice. Your LPO company should do the same for you.

Each page should revolve around a conversion task. Three major role types are visiting the accounting services site. There are local business owners, freelance creatives and middle aged homeowners. The conversion task for each is to gather their contact information for a free quote, but there are many paths to that conversion based on the needs, interests and triggers of those groups.

Your LPO company should determine a series of segments for your website and define specific tasks that those segments need to accomplish. The design, layout and content of your website will be determined by your segments. From there your LPO company should be able to plan and implement your strategy.

RED FLAGS FOR STRATEGY DEVELOPMENT

+ **Strategy development based on general user centered design principles:** User centered design principles are a good basis for landing page optimization, but a professional LPO company should use a customized strategy based on your market, your customers and the roles of the people who visit your site.

+ **No help with defining specific tasks for different roles:** Different types of users will need to accomplish different tasks on your website. Your LPO company should either incorporate several segmented landing pages or work to meet the needs of a small group of roles on one page. If your landing page is tightly targeted for one type of user, the LPO company should use persona marketing so the page speaks directly to the ideal user.

+ **Lack of defining for goals for strategy development:** Your LPO company should have a strategy that is targeted for specific goals. It is impossible to define conversion rates without a clear goal in mind. The goals for strategy development should be established at the offset.

QUESTIONS TO ASK ABOUT STRATEGY DEVELOPMENT

✓ **Will you be developing a persona for our landing page optimization?** Persona marketing is effective if your landing pages are tightly tied to a specific action and a specific group. A professional LPO company will be familiar with persona marketing and utilize it if it makes sense for your website and landing page.

✓ **How will you decide on tasks for my users?** The tasks should be defined by what your company wants to accomplish with the users, but your LPO company may also have suggestions based on your previous traffic figures and what they discover about your typical user types. For example, you may have been trying to convert a market of inexperienced Internet users to fill out a long form when you'll actually be more successful in educating them on your product and having them explore your site.

✓ **How will you use segment optimization in order to improve my conversions?** Your LPO company should be familiar with this term and utilize it if your website needs to appeal to a wide variety of role types.

IMPLEMENTATION

The implementation phase will take the information your LPO company gathered in the last phase and then use it to make specific changes to your website that will encourage the ideal users to take the ideal tasks in a shorter amount of time. Implementation will be custom tailored to your website and your users so it is difficult to say exactly what this part of the process will entail.

However, there are some basic strategies of user centered design that you should look for in your LPO company's implementation of optimization. There are three main messages that your landing page should convey with its various elements – trust, clarity and affinity.

o   Increasing trust will help your visitors trust the information, product or service you are offering. Without trust, they will be less likely to take the action. Your LPO company should use optimization techniques to minimize anxiety and increase trust.

o   Increasing clarity will make the desired action the logical choice for the visitor. Too much confusion on the page will turn off the visitor and will make it more likely for them to click away. An LPO company should minimize confusion and carve out a clear path for the visitor to take on your landing page.

o   Affinity is a sense of belonging and understanding. Although this may seem difficult to obtain from a website, it is possible if your LPO company implements specific changes in the tone and text of your website. Targeting language toward the visitor role and experience level will leave them feeling less alienated. People like to be understood and an LPO company that strives for affinity with visitors will make your landing page more successful.

(Ash)

Following are the main characteristics of usable design that your LPO company should use to make your site clear, familiar and trustworthy. An LPO company may provide these services in house, or they may recommend that you use a third party service to implement these changes. If they recommend a third party service, be sure to use the criteria and recommendations in other parts of this book in order to hire professionals in those fields.

GRAPHICS AND VISUAL DESIGN

The graphics and visual design of your landing page are the first impression that your website visitors encounter. With the right visual design you can immediately capture your audience's interest and get them to stay on the page to take the required action. In a split second, your website visitors are responding emotionally to your site and deciding whether they will stay.

Visual presentation is important because your visitors have a purely emotional response to design. If the visual presentation doesn't speak to them on the most base of levels, it isn't going to work. There are several elements that go into the visual presentation of your website. The page layout, graphics and colors will all work together in order to present an appealing visual image to your target market.

Page layout is the way that information is presented on your landing page. Complicated and convoluted page design reduces the clarity of the page and may reduce trust as well. The layout should guide your visitors toward the ideal action without asking too much of them. Never make them enter personal information before routing them to the information they are looking for. Never require them to scroll down the page to find the download they were looking for. The more steps you make them take between arriving on your page and getting their desired information, the more likely they will click away.

For example, a product based page layout will include the image of the product in the first screen shot as well as the price. In most cases, this is exactly what the visitor is looking for – what the item looks like and how much it costs. It would be counterintuitive for the page layout to include a large company shipping policy first or require the user to enter their name and email address before seeing the price of the item.

Graphics should be easy to load for a variety of different users. Unless your company is in a tech field, your LPO company should encourage you to use simple graphics that can be seen on the vast majority of computers, Your graphics should reflect the rest of your offline or online marketing materials to create a sense of cohesion for your company. This increases trust and affinity because your market will identify your company with the same presence across many different mediums.

In some cases, your LPO company may remove some of the graphics on your website to make the purposes of the landing page more clear. Graphics should be used to relate directly to the content of the landing page and not exist simply for show. Another important graphics element to include is any signifiers that indicate that you are a legitimate company – credit card logos, security insignias or stamps of approval from any certification boards in your industry will help increase trust and identify your legitimacy to users.

Color speaks volumes and can be a big factor in how much time your visitors spend on your landing page. Although the colors of your company's logo and brand identity are generally already selected, your LPO should advise on the best way to use them on your website to increase clarity, affinity and trust.

INFORMATION ARCHITECTURE

This is closely related to page layout but it also deals with the content that is displayed on your landing page. Although many company owners assume that the users will follow one path to achieve the desired task, the Internet opens itself up to a variety of different methods. Your LPO company should formulate several different plans within the same page to get the users to take the specific action.

For example, placing an opt in box for a newsletter in the upper right hand side bar as well as after a few paragraphs of text in the center area will appeal to users who want to register immediately versus those who need a bit more convincing. Keeping the opt in box in just one location will leave out one of the groups and decrease your conversion rate. Other examples of factors in information architecture include the presentation of cross-selling, ad space and other offers.

Your LPO company will likely build upon your existing results from your page layout and combine those results with their knowledge of proper information architecture for your niche and your visitor's needs. The entire structure of the information should be there to support the visitor and your desired task. Hiring a professional LPO company will help you see areas where you may be making assumptions about how your visitors will behave that are actually decreasing your conversions.

LANGUAGE AND TEXT

The third major element that is used to optimize your landing page is the language and text. This category includes headlines, content, anchor text for links and any other words that appear on the page. Keep in mind that website viewers have short attention spans. The words on your landing page have a lot to accomplish…and they need to do it quickly! Your LPO company will likely rework most of the text on your site since this is one area where companies tend to have the most trouble. If your LPO company does not handle copywriting in house, they will likely make a suggestion for an experienced copywriter to handle this job.

The language makeover starts with the structure of the text. We've discussed page layout and information architecture, but the structure of the text deals with how the information is conveyed without the pre-defined "text spaces." Your page layout and information architecture structure decide where the text goes and the text structure decides how the text is presented.

General rules for text structure on website pages dictate that the pages should be short. Short paragraphs, short sentences and bulleted listed are preferred to large blocks of text. Too much text is overwhelming from the visitor's perspective. In addition, your LPO company should organize your content in an inverted pyramid structure, similar to how newspaper articles are structured. The most important information should be conveyed at the top of the content with the less important content toward the bottom. This way the crux of the information will be read by your visitors as long as they read the first few paragraphs of your page.

Tone is also something that your LPO company should be adjusting in their optimization process. The tone of your website can make all the difference in their actions, especially if you are in a market where the competition is using a lot of high pressure sales copy and techniques. The tone of your website is the language used to create affinity and let the visitors know you understand their needs and are there to help. For example, the tone on an organic foods website would be written in a friendly way that invites the visitor to try the product. The tone on a financial planner's website would be more serious and to the point. The writing should be factual, oriented toward your specific task and be able to convey a lot of meaning in a small amount of space.

The LPO company should use these three elements together in order to optimize your landing page content and implement the strategies to meet the needs of your visitors and the ideal action. Graphics/visual design, information architecture and language/text are key areas in the effectiveness of your landing page. These are the tools through which all of the previous plans will be conveyed.

RED FLAGS FOR IMPLEMENTATION

+ **Leaving your company alone in the implementation process:** Not all LPO companies will have graphic designers, web designers and copywriters in house. If they do not, they should advise you on hiring these professionals and convey the framework that these professionals should use to optimize your landing page. If they don't give this framework, you'll be starting from square one with these third party professionals. The professionals may have different ideas based on their own standards for design or text which may not match up with your company's optimization standards. Ideally, your LPO company will complete these functions in house or outsource this part of the work and supervise it's completion.

- **Emphasizing one mode of implementation over the other:** Your landing page likely needs changes in visual design, information architecture and copywriting in order to be effective. Your LPO company should give you widespread techniques for optimization that touches all three of these key areas. It will be up to your budget and your time constraints to choose if all or just one of the methods are used, but the LPO company should at least offer solutions for all three areas.

- **Suggesting visual design changes that do not fit with the rest of your marketing materials:** Your landing page should be cohesive with your existing logo, color choices and other elements of your company's visual statement. A professional LPO company will not insist that you change your logo or your company's colors in order to fit their optimization plans. They should be able to optimize within your existing framework.

- **Creating text that is not in line with your marketing approach:** There are various methods that a copywriting professional can use to get your point across. However, the tone and content of your website should be in line with your current marketing approach. Direct sales copywriting is very different from informational copywriting. Make sure your LPO company (or the content provider) is familiar with the existing approach of your marketing materials. For example, if your website uses an informal tone and has a low pressure sales technique, the landing page shouldn't have a bold red headline that screams out at the customer.

QUESTIONS TO ASK ABOUT IMPLEMENTATION

✓ **How will you decrease the time it takes for my visitors to complete a task or make a decision?** They should respond with a plan that includes the three key areas – visual design, information architecture and copywriting. If any of these areas is missing from your description, ask about them.

✓ **Can you show me examples of how you decreased decision time on previous clients' websites?** Case studies from previous clients will help you see how the LPO company approached the problem and give you confidence in their methods. Be wary of any company that cannot show case studies and examples of the implementation step.

✓ **Who do you use for graphic design/web design/copywriting?** This will let you know whether they handle the content in house or outsource it to another company. If they use another company, ask for their partner's company name and do some investigation on that company to see if the quality will meet your needs. If you are not satisfied with the quality, you can always hire your own third party provider to implement the suggestions.

✓ **I have a graphic designer/web designer/copywriter that I trust. Can I bring them on board for the implementation of your suggestions?** A professional LPO company may not be comfortable with this option, since it may affect the results of their implementation process. However, they should be open to meeting, virtually, with your third party person. If they don't have services in house, you may be providing them with a new reliable contact that they can use in the future.

TESTING

The process of landing page optimization begins with testing what is working and what is not, and the implementation of changes follows the same pattern. Your LPO company should establish what needs to be changed first and then decide how those elements will be tested. In order to get the best results, they should change too much at once. Your site may need a complete overhaul in order to be optimized, but changing things too rapidly all at once won't tell them any more about your visitor's behavior than you knew previously. Although the end goal is to have a website that is working optimally, you should realize that it may take several rounds of testing to get it just right.

An experienced LPO will be able to advise you on the optimization process in general, but even they won't be able to predict what will happen with your site once it goes live. It's only through changes and observation of how the market responds that your LPO company will be able to fine tune your optimization.

After developing a list of elements that needs to be changed on the website through the previous steps, they should prioritize the elements that need to be changed. These are commonly referred to as the "tuning areas" that will be changed, tested and enhanced through the LPO company's work with your site.

In addition to establishing the tuning areas, the LPO company will set goals for the landing page, based on the site analysis and their consultation with you. It is essential to define what "success" will mean for your particular landing page, because it differs so much from page to page. For some companies, a decrease in bounce rate is a victory while others are happy with their bounce rate and want to get more people opting into their email marketing list. In addition, defining what constitutes a success from the start will give you and your LPO company an understanding of their role in the process and your expectations of their service.

SPLIT TESTING VS. MULTIVARIATE TESTING

Your LPO company will implement your testing process in two very different ways. Split testing and multivariate testing are two different approaches that each have their own benefits and drawbacks. In order to understand how your LPO company will approach the testing phase, it will be important to be familiar with how these two forms of testing work.

SPLIT TESTING

Also referred to A/B testing, split testing pairs two versions of the same page against one another, with the only change being one element. For example, the entire page will be the same in both cases, but the graphic in the upper left hand corner will be changed. Other examples of split testing is putting two different headlines against one another or changing the colors on the page.

The advantage of split testing is that your LPO company can analyze the results much quicker – they know that the increase in conversion is due to that specific element. The software to conduct split testing is very reasonable (and some is free) so generally speaking a split test will be less expensive than a multivariate test. However, split testing can be cumbersome and time consuming if you have a lot of changes to make on the page. Each element requires its own test and that could put the optimization of your site on a very slow schedule.

MULTIVARIATE TESTING

Multivariate testing allows your LPO company to implement changes in a wide variety of areas on your website and test them all at once. For example, they may change the headline, graphics and call to action statement in one fell swoop and then track the results. Multivariate testing not only shows your LPO company which combination of changes are working the best, but it also displays which individual factors are contributing the most conversion success.

Multivariate testing needs to be well organized and well documented. Although many free or inexpensive analytics software testing programs can handle multivariate testing, your LPO company will likely used a more advanced paid tool, which will lead to higher prices for their service. It's also a more involved undertaking for your LPO company, which will also result in a higher price than split testing. However, the major benefit of multivariate testing is that it will help your company get to the "magic combination" for conversions much more quickly. The likelihood of your company seeing results and reaching the "success" terms that you've established is greatly increased.

Multivariate testing can also allow for segment testing which can help your company understand how certain groups of users interact with your website. Segmenting visitors into groups through different testing channels will allow your LPO company to analyze how groups of people respond to different elements depending on their stage in the buying process.

DEFINING TUNING AREAS

No matter what type of testing your LPO company chooses to use, the selection of tuning areas is going to be significant to your success. Multivariate testing allows the LPO company to change a number of factors at once and see what works, but it's important to select a few elements to work with at once. Even though multivariate testing reduces the time it takes to test pages with split testing, it still requires time and resources. By being organized and focused, your LPO company can place their efforts where it matters most.

Your LPO company should refer back to the site analysis process to determine what would matter most to your target market and then make changes in those areas. A high tech market will be more impressed initially with design elements with content. A consumer market may prefer better pictures as opposed to effective copywriting.

Your LPO company should choose the elements that are the easiest to change and that will have the most impact first, rather than trying to change everything at once. They should also make changes based on your goals. If you want more opt-in subscribers testing that includes changing the text or placement of the opt in box is going to get you closer to your conversion goal.

RED FLAGS FOR TESTING

- **No clear definition of success:** Without agreeing on what constitutes a success for the testing phase of your website, your LPO company and you may be operating under completely different assumptions. Be sure that they are clear on their parameters from the start.

- **No prioritization or selection of tuning elements:** It's a bad sign when your LPO company does not select a method for testing that prioritizes the tuning elements in the order in which they'll best help your site. Their testing should be done in an orderly fashion and help support your company's goals for the site.

- **Using split testing when there are a variety of different factors to test:** Split testing isn't the best option when your landing page needs many different variables tested. An experienced LPO company will be able to accurately conduct multivariate testing to rapidly increase your results and more your landing page forward in terms of effectiveness.

- **Not spending enough time testing:** Testing is an ongoing process and will help define optimization strategies for the future. Be wary of an LPO company that promises rapid results that seem too good to be true. Proper testing of website optimization can take months.

- **No segment testing:** If your LPO company is utilizing multivariate testing, they should implement testing for particular segments in order to better understand your website's traffic and customers.

QUESTIONS TO ASK ABOUT TRACKING

- ✓ **What type of software or program will you use to track and test the optimization changes?** Your LPO company should explain why they chose that particular software and highlight a few key advantages to working with that tracking software. In short, they should prove to you why their choice of software will help your optimization goals.

- ✓ **How will you determine where to start with testing?** The LPO company should describe that they will focus on the elements that will get your company closer to its goals. The optimization methods should start with the most important elements to reaching those goals.

- ✓ **Will you use split testing or multivariate testing?** If there is more than one area to be optimized, your LPO company should respond that they will be using multivariate testing for your landing page optimization.

✓ **Will you provide ongoing support to implement the strategies that are suggested from the testing?** Once your LPO company finishes their first round of testing, they will know more about how your site should function. They should provide you with ongoing support for implementing those changes and further optimizing your website.

REPORTING

Landing optimization results cannot be properly evaluated without reporting. The reporting should take place at least on a per month basis. The reports received from your landing page optimization company can come in many forms. They will be directly tied to the goals that you and the company set forth at the start of the process.

It's important the reports do two things – they keep you up to date on the individual metrics and they communicate further plans for the future. The first portion is the facts and figures of your landing page conversions. It will detail how many visitors are arriving on your site, what they are doing on your site and how quickly they are leaving. This is important as a point of comparison with the initial site readings. It will immediately show whether or not their optimization efforts have improved or decreased your results.

The second part of the report should include steps that the landing page optimization will take in the coming weeks in order to maximize your existing results. Your LPO company should use the monthly reports as an opportunity to compare the current results with the projected outcome and site goals. If the current results do not match the goals, they should outline which steps they'll be taking in the coming month to help meet those goals.

RED FLAGS FOR REPORTING

♦ **No protocols for reporting:** Part of setting up a contract with an LPO company should be reviewing their policy for reporting. The reporting protocol should be consistent and clear from the start.

♦ **Reports are data-centric with no recommendations:** As outline above, data is only part of the picture when it comes to monthly reporting. The LPO company should also make recommendations in their report for future steps.

QUESTIONS TO ASK ABOUT REPORTING

✓ **How does reporting factor into our work with you?** The LPO company should explain that reporting is an ongoing part of their work with you. Reports are important to keep your company abreast of new developments in your site's performance.

✓ **Can I see a sample client report?** Looking at a typical report that your LPO company has created for another client (with identifiers removed of course) will let you evaluate whether or not the company is addresses the two key areas of reporting – statistical data and recommendations.

CONCLUSION

Landing page optimization can help make any advertising dollars you spend on driving traffic to your website more valuable. When you convert more of your incoming visitors, your return on investment for that traffic will be higher. Finding the right LPO company using the criteria listed in this chapter will help your website get better results.

CHAPTER SEVEN:

CONTENT CREATION SERVICES

Content creation is at the core of a successful presence online. Without the right content, your marketing message will fall on deaf ears. Although there have been many strides in recent years with audio and video content online, for the most part written content is what drives the Internet. In fact, there is a saying that "content is king" online. No matter what your business is, content marketing should be part of it.

An expert content marketing firm can help make your website stand out online. The writing on your website is your 24-hour salesperson. If the writing is doing its job, your business will make more money and have a wider impact in your market.

Your content company can help improve your conversions, draw in more traffic and increase your search engine rankings. This chapter will discuss the fine points of effective content marketing and how to select a content creation firm that will make a difference in your business.

CONTENT MARKETING BASICS

At the very minimum, there are a few pieces of content that your website will need in order to be inviting to your visitor. These pages will introduce your company and let your visitor know how you can help them. They act as a digital brochure for your company and will help your website visitors understand what you have to offer. The home page, about page and contact page are the bare essentials of what needs to be on your site in order to look legitimate and be effective for your market.

In order to create these portions of your website, your content company should ask you detailed questions about your site's purpose and your company's goal. This content cannot be written effectively without this information. If possible, send your content company a copy of your official brochure or a press release on your company so that they know where you are coming from and what you want to emphasize.

In addition to the basic content pages on your website, there are also many other types of content that can be used to create a better experience for the user. When you think about it, a home page, about us page and contact page don't inspire your audience to take a lot of action. Your website can be a lot more than a digital brochure. It can enhance your experience with the target market and position your company above your competitors. Your content creation company can do all of this by using content marketing.

Content marketing is in stark contrast to traditional marketing which interrupts the user's experience. Direct mail and advertising are examples of traditional marketing materials. While these forms of marketing try to capture the attention of the reader by advertising directly to them, content marketing takes a different approach. With content marketing you need to deliver information that is independently valuable.

The goals of content marketing are to create:

o   Trust

o   Credibility

o   Authority

With the right content, your market will seek you out and use you as the "go to" resource for your industry. It's all about offering information and not placing a sales pitch in front of your audience. Your content creation company should strive to introduce important concepts in your niche to your target market while introducing benefits of your product or service.

This is not to say that content marketing isn't action oriented. It's just done on a different level than in other forms of media and writing. Blogging, Web 2.0 platforms and social media tools have all given the masses a way to voice their opinion. Your marketing messages are part of this massive amount of content online. Your content company must create a way to make your website relevant and important amidst all of this noise.

The content your hired company creates for you must:

o Engage the reader

o Increase search engine rankings and traffic

o Increase links from other sites

o Increase conversions from website visitors

This may seem like a tall order for a simple article or landing, but in the hands of a competent content team, this is totally possible. Many website owners pay close attention to the structure and function of their site without giving a second thought to the content that will be there. The fanciest web design in the world won't be enough to keep your visitors there so you can sell them on your product or service. This chapter will help you choose the right content creation and management company that will have these keys in mind. For additional assistance with selecting a content creation company, see the latest content company rankings from topseos.com at http://www.topseos.com/rankings-of-best-content-creation-companies.

GENERAL CONTENT CREATION RED FLAGS

● **Misunderstanding of the basics of quality content creation and content marketing:** Content creation is not simply putting together articles with keywords. A proper content creation company will understand the research process behind creating quality content.

● **Lack of a proper content structure:** Every website needs basic components to run successfully. Your content creation company should be familiar with the basic structure required and make suggestions for covering these areas.

+ **Lack of quality content creation:** Samples from the content creation company clue you into whether they are able to build trust, credibility and authority with your website visitors.

+ **Lack of research:** Quality content creation requires in depth research on your company and the purpose of the content. Your content company cannot work with just a list of keywords. They should ask you multiple questions about your desires for the content, it's purpose, etc.

GENERAL QUESTIONS TO ASK ABOUT CONTENT CREATION

✓ **How long have you been creating content?** The more experience a content creation company is, the more likely that they will be able to research and write the type of content that you need.

✓ **What experience do you have writing content in my niche?** Finding a writer that has dealt with topics in your niche before will greatly improve their chances of being able to deliver the type of content you are looking for.

✓ **What type of content do you specialize in?** There are many varieties of web content services that your website and online marketing campaigns may need. Identify what you'll need and be sure you are getting services from a company that has experience with those services.

✓ **What are your policies for editing and rewrites?** Content creation is a collaborative process that may require a few rounds of edits. A professional content creation company will work at least two rounds of edits into their process so you are assured that you will be satisfied with the final product.

THE COMPONENTS OF QUALITY CONTENT

There are several components that work together to create quality content. Understanding these criteria will help you evaluate a content company and determine whether or not they have your best interests in mind. Look for these qualities in a content company and you'll know that they are handling your content needs correctly.

SEO FRIENDLY CONTENT

SEO, search engine optimization, is one of the ways that you can ensure your website is getting traffic from the right sources. Content is a large part of the equation. For more on SEO, please see the SEO chapter. For the purposes of this chapter, you need to know that SEO friendly content includes the right balance of keywords. Although traditionally SEO writing has been focused on including keywords in the text of the website, search engines are getting smarter in the way that they evaluate content. Keywords are still very important to search engine rankings but search engine algorithms take into account how others think about the content. This is evaluated by the words that others use to link to your content. Without quality content, you won't obtain valuable backlinks.

When your content company begins working on your website, they'll generally do a site analysis to determine what is necessary to get your site's content up to speed. They may do keyword research, or you can provide keywords based on your own research. Most websites either have too little content or too much content of the wrong type. Your content creation company will strive to incorporate the right type of content that will increase your search engine rankings and traffic.

Your chances of being found for related keywords in search engine rankings are greatly increased by using content that relates to those keywords. Even if you consider your website to be primarily a catalog of your products, you should still use content to draw in search engine traffic. Pictures and graphics don't bring in traffic as effectively as written text does. Even with the correct HTML tags and formatting, pictures and graphics won't rank higher than your website content. Your website content must include keywords in order to be effective for SEO

Search engine spiders are designed to crawl through websites in a way that mimics the way that humans read the website. The things that a human is more likely to consider important – like the text – is given higher weight by the search engine spiders. This means that the META information on your website isn't going to add up to higher search engine rankings unless there is quality content there to back it up. Here are a few ways that your content company can ensure that your content is SEO friendly and will draw in the traffic that you are looking for.

WRITING CONTENT THAT RELATES TO THE TOPIC

When search engine spiders first began ranking content years ago, they simply looked for the use of keywords and not how they operated within the framework of the website's topic. These days, they can recognize if the content is not in line with what the website is really about. Your content creation company should be creating content that matches closely with the keywords that you wish to be optimized for.

What this means is that if the keywords for your website are "improving memory" the content on your website should actually relate to "improving memory". It shouldn't just have the keyword "improving memory" thrown into the mix.

Every keyword has related keyword terms that will generally be found in the same content. For example, an article on "improving memory" will probably include references to "bad memory", "forgetting things" or "ginko biloba" (an herbal remedy that is said to improve memory). The search engine spider programs will look for use of these related terms in the content before they rank the content as being relevant to particular keywords. They give their stamp of approval once they see the related keywords. There are dozens of related keywords for any given keyword, so don't worry about your content company hitting the right ones exactly.

The easiest way for a content company to pass through these filters is to write naturally. Related keywords are naturally connected to your main keyword, so by writing a quality article the article will naturally become ranked for that keyword. Long gone are the days of keyword stuffed articles or pages of content that simply list keywords instead of providing real value. Search engine spiders have learned to filter out this type of content and are looking for content that appeal to human readers.

Several years ago, content was supposed to be written to meet a particular keyword density, like 3% or 5%. However, it is best to use natural keyword usage like what is described in this section. If your content company insists on using keyword density as a measure of keyword usage, it should be no more than 1% to 2%.

PROPER USE OF KEYWORDS

Even with the emphasis on natural writing, your content company can still optimize your content for specific keywords. They can do so through the following methods. These methods will ensure that your copy is optimized without appearing too "stuffed" with keywords.

o *Keyword distribution* – Your content company should use your keywords several times in the first paragraph, a few times throughout the body of the copy and once in the conclusion. This distribution will ensure that your content is optimized but that the use of the keywords is spread out. They shouldn't use synonyms for your keyword until well into the body of the content.

o *Using links as part of the content* – Your content company can use your keywords as anchor text for links within the body of the content. The link paired with your keyword places extra emphasis on that keyword for the search engine spiders.

o *Using headings and subheadings using the keywords* – Subheadings and headings are given extra importance by search engine spiders. By using H1 and H2 tags (and other tags) throughout your content, your content company can ensure that your keywords are getting extra emphasis. Bold and italicized phrases can also serve the same purpose.

SEO-FRIENDLY STRUCTURE

Your content creation company can structure your content in such a way so it makes your whole website more SEO friendly. Although the main structure of the website is normally up to the web designer, there are structural choices that the content company can make that will increase your search engine rankings. We've already discussed the importance of using keyword rich headings and subheadings. There are also other methods your content company can use to boost the SEO of the content.

One of the easiest ways to create an SEO friendly structure is to create a list article. List articles have the added benefit of being very easy for human eyes to read (which we'll discuss in the next section). Lists are important for SEO because they allow your writers to repeat keyword phrases often without that repetition appearing strange.

For example, if your golf website needs some content optimized for "golf swing tip" your writer could very easily create an optimized article by using the list structure. The list could have several entries titled "Golf Swing Tip" and each one would get a number (Golf Swing Tip #1, Golf Swing Tip #2). With ten tips, your article would have ten uses of your target keyword without appearing strange.

Another structure that smart content companies use to boost SEO is the how to article. If your website is service based or you need to have how to instructions for your products, these articles are a perfect fit. They combine quality content with the ability to repeat keywords several times. Frequently asked questions articles and definitions articles are also effective.

RED FLAGS FOR SEO FRIENDLY CONTENT

+ **Overemphasis on keywords:** Whether it's through stuffing keywords in content or planning content strictly based on keywords, your content company can do your website a disservice by emphasizing keywords too much. Keywords are important for SEO purposes but search engine spiders have evolved to the point where they can tell if you are trying too hard to optimize a page for a keyword.

+ **Ignoring keywords completely:** On the other side of the fence, keywords need to be the foundation of most of your content on your website and ignoring them can mean you're missing out on a great deal of traffic that you could be getting otherwise. Your content company should develop your content based on your keywords with some extra content added that is based on the needs of the market.

QUESTION TO ASK ABOUT SEO CONTENT

✓ **What is your process for developing content with my company?** This is a good general question to ask, whether you want SEO content or another type of content. The content creation company should respond with a specific process for evaluating your existing website content, developing a content strategy and communicating strict deadlines for delivery of that content.

✓ **Can I see a sample of your SEO Content? What keyword was this content optimized for?** This is an essential question because without a sample you cannot ensure that the company will be able to write well. By knowing which keyword the content was optimized for, you can evaluate whether or not the company is putting their content together in a manner that will be pleasing to search engine spiders and human readers.

✓ **How do you select the keywords?** There are many right answers to this question. They may ask you for the keyword list or provide their own. If they provide their own, ask about their methods of determining the right keywords for your website. Look for sound keyword research methods, like those described in the SEO chapter. If the company provides SEO services as well as content writing, their keyword selection methods should be excellent.

✓ **How will you optimize my content for these keywords?** Even if they have shown you a sample, this question helps you analyze their thought process in creating that sample. They should explain methods similar to what was discussed in this section including keyword placement, related keywords, structure and using headings.

✓ **What kind of keyword density do you use?** This is a trick question. A professional content company should explain to you that natural keyword usage is preferably to keyword density measurements.

Case Study #1: Masala Foods

Content Company: Tengo Communications (http://www.tengocommunications.com/)

SEO Friendly Content: Masala Foods is a gourmet caterer of healthy Indian Food. They approached Tengo Communications for assistance with search marketing optimization and branding through content. Tengo Communications created copy for the company's website that incorporated important keywords in order to increase search engine rankings.

Case Study #2: eDrugstore.md

Content Company: The Write Content (http://www.thewritecontent.com)

SEO Friendly Content: eDrugstore.com is a provider of Online Pharmaceutical Sales. In order to compete with some of the more established online pharmacies, they enlisted the services of The Write Content. In order to attract a steady stream of new visitors and cut cost on PPC advertising, the Write Content rewrote eDrugstore.md's detailed product description pages with keyword optimization methods.

MARKETABLE CONTENT

There are two types of readers that will take in your website content – the search engine spiders and the website visitors. Keyword requirements and article structure are meant to attract search engine spiders but the other half of the equation can only be persuaded with quality content. When your content company creates marketable content, they are assisting your website with becoming an authority in your niche. Having an authority site means that you'll get additional backlinks to your website which can in turn lead to higher search engine rankings. Quality content that engages the reader will also help build trust with your audience and can increase your conversions.

The Internet is full of information of which your website is only a miniscule piece. Thousands of other websites are competing with your website for your reader's attention. Your content company's goal should be to hold onto your reader's attention long enough so you can deliver your marketing message and build trust.

Website writing is frequently criticized as being less complex and engaging than print media. Even though there is a slight emphasis on keywords and the technical aspects of optimization, your content company should understand the needs of the human reader. Excellent quality writing with keywords incorporated will always do better overtime than content that has been falsely propped up by SEO methods.

MEETING THE NEEDS OF THE AUDIENCE

Keywords help content writers meet the needs of the search engine but to create really excellent content, your content company should strive to meet the needs of the audience. They should treat each keyword as if it were a piece of the puzzle. While keyword research shows how many people are looking for a particular topic, it's your writer's job to display the information that people are actually looking for when they enter that phrase into the search engines.

People come to websites to answer a question or complete a task. When they type "cure for shingles" into the search engine it is pretty obvious what they are looking for. However, words like "snow leopards" or "investing" may not be as obvious. A quality content company will determine what is best for the reader in the case of these broad keywords.

Staying on topic is not the same as engaging the reader. When you look for information on "cruises to Mexico" you probably aren't looking for a detailed history of how cruise lines to Mexico were established or how to get a job on a cruise line to Mexico. Although articles on either of those topics would still be considered "on topic" they are far from useful to your reader. Good content companies understand this difference and their writers will be able to produce on topic content that is actually helpful.

STYLISTIC CHOICES

In addition to meeting the needs of the reader through the right approach to the topic, a content company can also use style to engage the reader and keep them focused on your website. Here are a few different methods that can be employed.

o   *Writing in a conversational tone:* Formal writing is hard to read online. In addition, reading online is a voluntary activity and nobody wants to spend their time focusing in on boring content. Even on rather complicated topics, a conversational tone works best for online readers. Print magazine articles are written in a style very similar to what is most successful online.

o   *Using personal pronouns:* "You", "yours", and "we" are very rarely used in other forms of writing. However, reading on the web is a personal experience so your content company should use personal pronouns to engage the reader.

o   *Using short sentences and short paragraphs:* Have you ever encountered a large block of text online and quickly clicked away? Your website visitors will have the same experience if your website doesn't follow this guideline. Short sentences and short paragraphs are easier to read online. Computer screens are difficult to focus in on for long periods of time so people normally scan through the text online looking for headings, bold text and short "chunks" of information. Your content company should write in an easily "digestible" format and make use of short sentences and paragraphs to keep the reader engaged.

o   *Treating each page as a landing page:* The chance of your website visitor coming to your website through your homepage is very rare unless you are only building links to your homepage (which isn't a sound link building strategy – see the Link Building chapter for more information on this topic). Most SEO professionals recommend that your links should be built to the various inner pages of your website. This means that there is a high likelihood that one of your interior pages will be your visitor's first introduction to your website. For this reason, your content company should avoid referring to other pages frequently and the independent pages shouldn't build upon one another. Your content creation company should also make use of links between pages in order to guide the user through the different elements of the website.

TYPES OF ARTICLES

The vast majority of the content that will be created on your behalf will be articles. Articles are perfect for website content and they can be used in link building (through article marketing) and for social networking (by distributing the on Web 2.0 sites). There are other types of writing that your content company can provided, which will be covered in a later section.

Typically articles are 400 to 500 words long, which is a good size to quickly skim through in one sitting. Since website visitors skim rather than read articles are kept short so that they serve the interests of your readers. Excellent content writing companies will make each article on your website a wealth of information. They will not have the same approach to the topic, although they will come from the same angle. That is to say that your content company should provide you with articles that approach the keywords in different ways but still come from the same point of view.

For example, you give your content creation company the keywords "organic garden mulch," "organic herb garden," "organic vegetable garden," and "organic tomato plant." Your content company should not come back with a list of articles that looks like this:

"How to Make Organic Garden Mulch"
"How to Make an Organic Herb Garden"
"How to Make an Organic Vegetable Garden"
"How to Grow an Organic Tomato Plant"

This isn't enough variety to please the average reader. The articles should be of varying types. Having all "how to" articles or all "question and answer" articles will make your site one dimensional and flat. In addition, the articles shouldn't contradict one another. Your content company shouldn't deliver an article on the benefits or an organic vegetable garden and an article about how hard it is to grow organic tomato plants.

Other types of articles your content company should use include:

o   Problem articles – This type of article will illustrate a common problem that people in the market have with a topic. For example "Confused about Organic Mulch? You're Not Alone."

o   Tips list articles – This article is great for SEO purposes but also helps readers because it has easily digestible content. For example "3 Must Have Tips for a Successful Organic Herb Garden." Other list types includes reasons lists ("3 Reasons to Grow Organic Tomatoes"), benefits lists ("Why You Should be Making Organic Mulch") and mistakes lists ("The Top 5 Mistakes People Make with Organic Vegetable Gardens").

o   Frequently asked questions articles – This type of article is easy for website visitors to read because it is organized in bite sized chunks of information. An example of this type of article is "Frequently Asked Questions about Making Organic Mulch."

A professional content creation company will use these article types and several more to give your website a variety of content that will engage readers. Insist that your content company plan a variety of different articles to both meet your keyword needs and the needs of your website visitors.

CONTENT STRATEGY

Your involvement with your content company depends on many factors. Some people may be comfortable with having the content company deliver helpful articles and then having their web designer upload the articles. Others may know what they want as far as content goes but aren't comfortable with their writing skills. Still others may want a full service content provider that can help develop the writing on their website for a long time to come.

If you're in the third group, you need to look for a content creation company that has experience in developing a content strategy. A content strategy is a long term approach to content creation that goes beyond simply writing articles and putting them on a website. Experienced content strategists will be able to help you determine how your website should appear now and how it will grow in the future.

Content strategy deals with planning out and managing content through its complete lifecycle on your website. The strategy is based on your website's needs and your company's place within the larger marketplace. It includes evaluation of the current content, a plan for implementation and the management of the plan in the future.

In order to attract new visitors and increase their importance in the niche, websites need to consistently have new content. A content company that has experience with content strategy will be able to plan out this new content and help you implement it on the site. It's the ultimate form of making the content marketable because it is based on the market's changing needs. It positions your website as an authority.

RED FLAGS FOR MARKETABLE CONTENT

+ **Not understanding the difference between marketable content and keyword driven content:** Look at the samples that your content company gives you in addition to their website. Look for evidence of their understanding the needs of the reader.

+ **No development stage:** Even if your content company does not offer content strategy services, there should be a development stage of content creation. Your content company should ask you about your ideal customer and the most common questions that those customers have in relation to the keywords that the writers will be using. They will also ask you about your point of view for a website. There's a big difference between an article on "gun control" for a website that supports second amendment rights and an article on "gun control" for a website that wants to limit gun usage.

+ **The same type of article again and again:** The articles provided by your content company should give your website visitors a variety of perspectives on your topic. They should be thematically different but express the same viewpoint. On your first project together, ask for a list of possible article titles for your website or suggest article titles to make sure that there is variety on the site.

QUESTIONS TO ASK ABOUT MARKETABLE CONTENT

✓ **How do you make sure that my website visitors will be interested in the content?** They should describe writing for the reader and search engines at the same time. The content company should also emphasize that their content is driven by answering questions and providing helpful, accurate information.

✓ **Do you provide content strategy assistance?** This is not a requirement but it is good to know. If they do, ask how they will determine the strategy for your website. They should take a look at other websites in the industry and determine the type of content that should be used on your website and in other content formats.

✓ **Do you outsource your writing? Where are they outsourced from?** If the content creation company is not doing the writing in house, you need to know where the writing is coming from. Cultural differences and poor English skills can make a difference in the ability for the content to be marketable.

✓ **Give me an idea of the type of article you'd write for this keyword:** _____ This strategy will help you determine whether a content company can understand the distinction between writing to meet the topic and writing to attract the website visitor. Samples can show you this as well, but asking this question will help you understand their thought process. Unlike the question in the previous section, you're looking for the style of the article and not the keyword optimization.

Case Study #1: Masala Foods

Marketable Content: Tengo Communications utilized a narrative throughout their content for Masala Foods that enticed readers while substantiating the business and communicating company branding messages.

Case Study #2: eDrugstore.md

Marketable Content: The Write Content created product description pages that informed and educated visitors on various products offered by the website. It also helped establish brand identity and build company credibility. In addition to editing product description pages, The Write Content implemented a content strategy that included general information articles. These articles positioned eDrugstore.md as an information hub for the market, rather than just an online store.

CONVERSION DRIVEN CONTENT

All the informational articles in the world won't make a difference if your website isn't helping you reach your goals. If your sales aren't increasing, you aren't receiving any new backlinks and your site isn't increasing in the search engine listings, your content isn't doing its job. Your content company should focus on creating conversion driven content that not only informs but encourages the visitor to take a specific action.

You should communicate to your content company what type of action you want the website visitor to take on that page. Each page may be different or you may have your entire website geared toward one purpose. Sample actions can include:

o   Filling out a form to get more information.

o   Visiting another part of the site.

o   Contacting you through email.

o   Purchasing a product.

Even the page in question is just designed to inform your website visitor about your business, it still has a conversion purpose – building trust. Your content company can increase your conversions with content in the following ways.

PRE-SELLING CONTENT

Even if there is no direct call to action within the text of your website, all of the content on your website should be "pre-selling" your website's purpose. We discussed earlier how none of the content should be against your point of view on the site. While this is true it's more accurate to say that all of the content on your site should be biased toward your point of view. It's not enough for your content to not support your opposing point of view; it has to subtly push your own agenda.

You've only got your reader for a short period of time. Your content company has to create content that will grab their attention and convince them of your point of view quickly. This needs to be done in a subtle and effective way. Your content company should be well-versed in pre-selling the audience on an idea.

What is pre-selling? Pre-selling is the act of building trust with your audience. Your content writer, given the right information, will create informational articles that will build trust. Building trust will help pre-sell the audience on buying from you. When people come looking for information online they generally don't want to be sold to. They have their guard up and are resistant to purchasing. With pre-selling content you can break down that resistance.

Giving away content for free is an act of building trust. If your website is full of quality, free information your audience will begin to see you as a resource and a generous person (or company). They've seen what you have to offer for free and they'll likely want to see what you have for sale.

The content writer can also help pre-sell the audience on your particular product or service. For example, your website sells eco-friendly roofing material. It's more expensive than standard roofing material, but the benefits to the environment are well worth it.

The content on your website should serve several purposes:

1. Inform people about the existence of eco-friendly roofing material.

2. Show people how traditional roofing material is harmful to the environment and more expensive in the long run.

3. Promote the benefits of eco-friendly roofing material.

4. Show the people how purchasing from your company is the best decision.

Some of this content will be presented in article form. Some of it may be presented as standard website content. Blog posts, social networking content and newsletters are also considered content. No matter what form it takes, all of it should be pre-selling your product or service.

Your content company should be committed to using pre-selling strategies for your website. When they pre-sell they make your target market more comfortable with your message and they understand that you aren't just in it to make the sale. Your content company can help you convey this attitude on your website. This will help convert website traffic into buyers and buyers into lifelong customers.

LINK BAIT ARTICLES

Your content company should create articles that encourage the likelihood of quality backlinks. Backlinks are vital to the search engine rankings of your site. For more about the importance of backlinks, see the link building chapter. Link bait articles are designed to encourage others to link to your website. With the increase in social media sites like Digg.com and StumbleUpon.com, individual pages on your website may get enough attention to increase your traffic and your website listing. A smart content creation company will not only look to create value to existing website visitors but will work to draw in traffic with individual articles.

Link bait articles are different from the standard content that a content creation company will put on your website. They are most often delivered in blog format to increase social bookmarking and community discussion. This is one of the main reasons that you may want to consider developing a blog for your company's website, with the help of a content creation company. We will go over additional reasons for having a blog in the next section.

Link bait is content that has been created with the sole purposes of encouraging backlinks. This can be done through usefulness, controversy or other means. For example, a content creation company could create several different link bait type articles for a car rental company. They could create an article that lists the top 10 makes and models of cars that are likely to be pulled over by cops (useful content). They could create an article that attacks the car rental industry in general for several practices that other car rental companies use (controversial).

Link bait should comprise 10 to 20 percent of the overall content on your website. Although link bait is effective in boosting traffic it often does not contain the kind of information that a prospect needs to make a decision on taking the next step with your company. However, it does get the right type of traffic to your website and can increase your search engine rankings overall. Link bait should be a part of the conversion based strategies that your content creation company uses.

MULTIPLE CALLS TO ACTION

Each page on your website is an opportunity for your visitor to take the next step with your company, whether that is filling out a form, making a purchase or calling your company. Your content creation company can formulate the content in such a way that there will be multiple closes throughout your website, for increased conversions. Your content creation company should be adding calls to action at the end of each article, blog post or web page. The call to action should relate directly to the content that precedes it. This means that your content creation company should not slap the same call to action statement at the end of each web page.

For example, your website is designed to sell your golf equipment. Each piece of content on that website is optimized for a particular piece of golf equipment. Therefore, each call to action on each page should also relate to that particular piece of golf equipment. An article on Mark VIII golf tees should end with "Get your own Mark VIII golf tees and shave points off of your par today!" and not something generic. Although the body of the article shouldn't be promotional, your content creation company should understand that a strong call to action at the end of the copy will make all the difference in the effectiveness of the content.

RED FLAGS FOR CONVERSION DRIVEN CONTENT

- **Lack of focus on conversion:** Unskilled content writers will take a keyword and create an article around it without any thought as to how that article will be perceived by the reader. For example, an article with the keyword "paper recycling" that focuses on the air pollution caused by paper recycling plants will be of no use to a company that touts recycled paper gifts as an earth-friendly alternative. Before creating any content, your content creation team should ask you about your goals for your site so the content can match its purpose.

- **No calls to action:** Even with conversion focused content, there can still be a problem if the content does not have a clear call to action at the end of each page. Your content writing staff should be able to incorporate a subtle call to action statement at the end of every piece of content they produce.

♦ **Overly promotional content:** Your website should sell through persuasive and informative content. It should not scream out at the reader to BUY NOW!, unless there is a call to action being used at the end of the content. Conversion driven content should follow the characteristics set out before and not be overly promotional in nature. The call to action is a different matter entirely.

QUESTIONS TO ASK ABOUT CONVERSION DRIVEN CONTENT

✓ **How do you create content that pre-sells?** The content company should respond with an answer that implies that they are familiar with pre-selling. They should focus on the benefits of your product or service and explain how they'll create content to emphasize these benefits and emphasize the need for your product or service.

✓ **What kind of link bait articles can you create for me?** This question tests their familiarity with link bait and their ability to come up with fresh, original ideas. Look for a quick response or experience in your niche.

✓ **How do you work calls to action into your content?** The content creation company should be able to use calls to action within regular content. If they respond that their content is for informational purposes only, they don't have sufficient experience with content marketing. Many writers only focus on informative articles and have no experience with creating content that also pre-sells. Look for a company that can provide you with both.

Case Study #1: Masala Foods

Conversion Driven Content: Tengo Communications utilized the principles of conversion driven content in order to utilize the Masala Foods website as a platform for attracting visitors and converting them into buyers.

Case Study #2: eDrugstore.md

Conversion Driven Content: The Write Content used persuasion copywriting techniques in order to help the website sell more products.

VARIANCE

A content marketing company should have the ability to provide you with several different types of writing styles. As alluded to before, there are many types of writing used online. Website content, article marketing, blogs and email marketing are all types of content that will be provided by quality content marketing companies.

Variance with content is essential because the content of your site will serve several different purposes. Earlier in this chapter, we discussed many different types of articles that you can have created for your website (how to, list articles, tips articles, etc). However, there are also many different levels of writing and types of writing that a content creation company can provide you. Look for the following types of writing available from your content creation company. You may not need them all immediately, but it is better to hire a company that has widespread capabilities.

WEBSITE CONTENT

As previously discussed, website content is static content that is always present on your website. It is sometimes referred to as "cornerstone content." It is designed to draw the reader into the site and pre-sell them on your product or service. Your website content should be high quality and centered on competitive keywords.

A good content creation company will be able to create website content that is appealing to human readers and search engine spiders alike. It will be impressive in quality and will encourage other websites to link back to it. Good content will be targeted toward the reader with their interests in mind. It will succinctly explain the purpose of your website and prove to the reader why your company should earn their business.

Depending on your business model, there are many different types of website content that can be used on your website. They include:

o   SEO Articles – articles that are optimized for specific keyword terms in order to increase the search engine ranking of your website for those terms. Short keyword rich articles, longer informative content and optimized landing pages comprise the type of SEO articles that a company can create for your website.

o   Product descriptions – Product descriptions can be optimized for keywords and designed to pre-sell website visitors on the products on your site. They can make an impact just in a few paragraphs

BLOG ARTICLES

Blogs are an important tool for interacting with your target market. Like website content, they offer you the ability to inform and persuade your target market. However, they are not written in the same way. In order to stay relevant, blogs need to be updated consistently, at least once per week and as much as once per day. A content creation company that is familiar with blogging will understand the stylistic differences between website content and blog content.

Simply put, blog content is more engaging and more involved with the users. Comments are normally allowed on blogs so readers can give feedback on the topic and add their own opinion. This means that the content should lend itself to discussion. Blogs are a good platform for link bait content since people are often more apt to link to blog posts than static web content. Blogs can also incorporate special widgets will make it easy for web visitors to add social bookmarks to blog posts.

Blog content should also be relevant to the time of year they are posted. Unlike website content, which must stay evergreen, blog content can be on target for seasonal topics. Your gardening website can give seasonal tips on gardening in spring, summer, fall and winter, which improves the user experience.

Search engines tend to favor blogs over static websites because they are consistently updated. A qualified content company can help you establish a blogging editorial calendar and create content that can be drip fed onto your blog for the coming weeks and months.

EMAIL MARKETING

Creating informational and persuasive email marketing messages is a different skill set from articles and blog posts. Many content creation companies offer email marketing messages, but some do not. Persuasive emails often fall under the category of "copywriting" and you may have better luck with an individual or company that specializes in sales letters and other direct response marketing writing. For more on email marketing messages, please see the Email marketing section of this book.

ARTICLE MARKETING

In addition to serving as website content, articles can also be used as a way to drive traffic back to a website. Article marketing is the process of creating and distributing articles to article directories. These directories display your article, along with a link back to your website, for free. Your article can be copied from the directory and used on someone else's site, as long as they include the link back to your website. Article marketing will help bring in traffic wherever the articles are displayed. It will also help increase search engine ranking because each link back to your website will increase your relevance for your selected keyword terms.

Article marketing is similar to creating website content, but each article must also be targeted to one specific keyword rather than several like website content. Articles for article marketing also require an interesting title that will encourage readers and publishers to click on your article and select it from the thousands of other articles in the directory.

Most content creation companies offer article distribution services as part of their writing services. If you choose this method of marketing, it is important that your content creation company distributes your articles to quality directories that have a high Google Page Rank (PR3 or above). Your search engine results rankings will be increased if high ranking websites link to you. For more on obtaining quality backlinks, see the link building chapter.

PRESS RELEASES

Press releases are an important part of creating backlinks to your website and spreading the word about your content. Your content company should be able to create a press release that is on a newsworthy subject and that uses specific keyword phrases to increase traffic from search engines.

Look for your content company to provide press release writing services as well as distribution to press release directories. For more on what makes effective press releases work, see the press release distribution chapter.

REPORT AND EBOOK CREATION

Reports and ebooks are longer pieces of content that can be used for promotion online. An industry report on a hot topic or an informative ebook given away for free on your website can build trust with your niche, increase your backlinks and improve your company's importance within the niche.

A common type of report is a white paper. White papers are authoritative reports that are targeted toward specific issues and how to solve them. Your content creation company will use white papers to educate the readers on a common problem that your company can solve. They are not promotional in nature but can be used as a marketing or sales tool by your company. A content creation company can also create workbooks that will help your target market see simple solutions to complex problems. For example, a financial services website can reach their market with a budgeting workbook.

Courses, delivered on the website, in report form or through email, are another type of long form content. Courses can teach the target market about the problem and lead them through step by step solutions, which include your company as a big part of the solution. Ebooks are longer than reports and help build your brand long after your visitors leave your website. Ebooks can also be published as in print books that help make your company an authority in your industry.

Reports and ebooks require a great deal of organization and in depth knowledge about a topic. If they are going to be part of your future online content marketing strategy, make sure that you hire a content creation company that has experience in your niche and experience with writing pieces of this length. Otherwise you may end up with a low quality report that is stuffed with "filler" content and unsophisticated writing.

WEB TRANSLATION

Translation services are an added bonus if you are working in an international market. Not all content creation companies offer translation services but it is convenient if they do. Creating foreign language content on mirror sites of your main site can help you have a larger impact in your market and stand head and shoulders above your competition. It makes sense to have the company that created your content offer the alternative versions of the content in a foreign language.

CONTENT CREATION AND SOCIAL MEDIA

Many areas of social media and content creation intermingle. Blogs are used as social media devices and the content that is used on wiki pages, Twitter accounts, Facebook accounts and other social media tools can be created by content creation companies. If you plan to use social media marketing in the future, it is helpful to find a content creation company that has experience in these areas. If the content creation company does not offer social media services, they should at least be familiar with how to create content for these platforms. It will save you time in the long run when you choose to incorporate social media into your business plans.

RED FLAGS FOR CONTENT VARIANCE

- **Samples include only one type of content:** Experienced content marketing companies should be able to show you a wide variety of samples and tell you the purpose of each of these samples. If their samples are only of informative articles, they may specialize in this type of writing and will not be able to meet your other content writing needs. If the company's website does not specify different types of writing, ask for additional samples or inquire about the specific type of writing you'd like to have.

- **Samples claim to be different types of content, but the writing style is the same:** Experienced content creation companies will understand the difference between a blog post and an ebook page. As you read the samples, ask yourself if they are meeting the style requirements of the particular type of writing. If you aren't familiar with the style requirements, read other examples of writing in the category you are looking for (for example, a well-written ebook or an excellent blog post).

QUESTIONS TO ASK ABOUT CONTENT VARIANCE

✓ **What types of writing services does your company offer?** Look for companies that answer this question with confidence and detail. Specific references to types of content (webpage content, articles for article marketing, link bait posts, etc) is preferable to an answer of "whatever you need." You can answer follow up questions on each type of content that interests your business.

✓ **How do you make the distinction between content for a website and the articles for article marketing?** An appropriate answer would be that while both are geared toward popular keywords, the content on the website is designed to pre-sell the audience on taking the next step with your company. The article marketing articles should have catchy titles, a developed bio box and be leading the reader toward clicking back to the main website.

✓ **Are you familiar with creating email marketing messages?** If the company says yes, be sure to ask for samples, open rates and conversion rates for those emails. Hiring a content creation company that does not have experience with email marketing to write emails for your website can cost you prospects.

✓ **Are you capable of creating ebook content that could be published?** If your company plans on creating an ebook, it makes sense to have it created by a company that can give you a quality product that can be published in print.

Case Study #1: Masala Foods

Content Variance: As part of their service to Masala Foods, Tengo Communications created print brochures and other marketing collateral.

Case Study #2: eDrugstore.md

Content Variance: In addition to product descriptions, The Write Content added information articles to the site to make it a content hub rather than just an ecommerce site. This increases the value for the visitor and will increase the number of backlinks coming to the site.

WRITING STANDARDS

Writing standards are important to your website's content. It's essential that your content creation company has professional writing experience and is familiar with various writing standards. Most commonly website writers will adhere to MLA or APA standards. There are also additional standards that are specific to the Internet. It is the job of your content creation company to adhere to these standards so your website and all related content are presented in a professional manner.

RED FLAGS FOR WRITING STANDARDS

♦ **Misspellings and grammatical errors on the content creation company's website:** This is a major sign that the company is looking to take your money and provide you with poor service. If a company can't take enough time to polish their own website, they won't do a good job with your own content.

♦ **Professional experience in offline writing:** Often low quality content creators will enter the field without having offline writing experience. A college degree, journalism experience and other professional offline writing experience will give a writer or a writing team familiarity with MLA and APA guidelines.

QUESTIONS TO ASK ABOUT WRITING STANDARDS

✓ **Does your team write within MLA or APA guidelines?** This question will let you know their level of familiarity and what you can expect when you work with the company with regards to writing standards.

✓ **If you outsource your content, what type of writing standards do you hold your writers to? Do you check for plagiarism?** Any content company that uses outsourced content should have writing standards for their workers. In addition, they should run the content through plagiarism checker, like Copyscape. com, to ensure that the work they are receiving is original.

CONTENT CREATION CASE STUDIES – FINAL RESULTS

Throughout this chapter, we've looked at examples from the content creation efforts of two companies, Tengo Communications and The Write Content. They used the five important areas of content creation outlined in this chapter – SEO friendly, marketable content, conversion driven content, variance and writing standards. Here are the results from their content creation efforts.

TENGO COMMUNICATIONS CASE STUDY FOR MASALA FOODS

As a result of their content efforts online and in print, Tengo Communications was able to help Masala clearly differentiate its offering from the competition. Their white label grocery store line, presence in Los Angeles farmer's markets and their catering business were all benefitted from a renewed content approach. High quality copy created by Tengo Communications placed their business at the top of the search engine rankings for key terms.

THE WRITE CONTENT CASE STUDY FOR EDRUGSTORE. MD

Working with The Write Content, founder and CEO Gabriel White discovered a way to build his eDrugstore brand, establish his website as a credible source of health information, and enhance his company's image. Traffic increased by 20% in the first month and another 10% in the following 15 days. The articles and content produced 300 new orders for the company immediately after The Write Content implemented their content strategy.

CONCLUSION

Content creation is becoming a more important method of marketing online as search engine spiders are able to evaluate the relevancy of the written word. Understanding this, it's important to enlist the help of an experienced and knowledgeable content creation company. There are multiple ways that writing can help your online marketing strategy, so it's essential that you find the right type of writer or writing team to help you.

CHAPTER EIGHT:

VIDEO SEO SERVICES

Multimedia content is on the rise online and a professional video SEO company can help you garner more traffic and position your company better within your field. Video marketing can be used to get attention for your company, explain product features, highlight customer testimonials and answer important questions. Virtually any topic that can be covered in written content can be adapted and distributed via video.

Video creation can run the gamut from screen capture videos to professionally produced live videos. When you hire a video SEO firm, you are employing them to help your videos get better search engine results. With a popular video, your traffic and leads will increase. These benefits are not likely without proper optimization of the video title, keywords and distribution channels. A professional video SEO company will be able to put your video in front of the right group of web visitors for the best results.

In this chapter, we'll go over why online video has become such an important part of marketing online. If video marketing is currently not a part of your Internet marketing plan, you may want to consider incorporating it. We'll also go over the exact duties of a video SEO firm and what you can expect from professional service in this area. topseos.

com maintains a list of video SEO companies that you can refer to at http://www.topseos.com/rankings-of-best-video-seo-companies.

WHY VIDEO MATTERS

A video SEO company will give your business the boost it needs in a vital and emerging area of online marketing. The use of Internet video has rapidly increased over the last five years. It has transformed from a format of pure entertainment to one that is more usable for business. With video marketing, you can reach your target consumers in new ways.

Video marketing matters because videos are more popular now than ever before. It's estimated that 30 hours of video are uploaded to YouTube every minute, which translates to 1,200 years of video uploaded every year. (Snow) YouTube has been joined by Revver.com, Metacafe.com, Dailymotion.com, Viddler.com and many other video sharing sites that offer your company the opportunity to market to the masses. In response, video specific search engines, like Truveo.com, have also been developed to help searchers find the videos that they are looking for.

The mass consumption of videos as a regular form of information is growing rapidly to meet rising demands for entertaining content. Video has the capacity to capture your audience's attention on a whole new level. Considering the fact that the average web user's attention span is very short, it's clear to see why video marketing is important. When you produce and distribute an entertaining video, your market is likely to stay on your website longer and the chances of them purchasing from you are increased.

Not only are there more videos than ever, search engines are taking notice and ranking videos in search engine listings. Google's "universal search" feature was implemented in May 2007. The feature integrates regular search engine results with news results and video results for the keyword term. This gives online marketers another avenue for ranking high for keyword terms. Most importantly, obtaining rankings with video is much more accessible.

Videos are 53 more times likely to appear within the first page of search results than a text page for the same keyword. (Gannes) Videos are dominant in universal search results. Over 38% of people who used Google were delivered relevant video results in 2008 (Robertson) and those numbers are growing.

In addition, results for videos appear within just a few days of submission, rather than weeks with text pages (Wayne). Producing and distributing a video can give your company a distinct advantage over the competition. Web searchers are seeking out video content, videos are more likely to be ranked than text and videos are garnering quicker results. If you are trying to target competitive keywords, your videos will get you quicker and better results than using traditional text page SEO.

There are, however, some drawbacks to video marketing that need to be considered. Search engine robots cannot read the content of the video so they cannot determine the quality of the video. Unlike written text where the search engines can analyze the quality of the content and rank it accordingly, videos are ranked based on the way they are labeled, tagged and linked to.

Recent developments have started to solve this problem. YouTube now allows users the ability to place captions on their videos. This means your video SEO company can add a transcript of the video, which can be indexed by Google and add value to your listing. In addition, SpeakerText allows your video SEO company the ability to quickly transcribe your video to text which is pasted onto your site. (Snow)

In addition, video sharing sites are highly user interactive, which means that the viewers of the videos are allowed to rate, tag and otherwise classify your videos. Although this user classification is completely dependent upon the users, your video SEO company can encourage proper tagging, links and ratings for your videos that will help increase their relevance for specific keyword terms.

With these advancements, the few drawbacks of video marketing are being solved. Video marketing can be an excellent addition to your online marketing efforts. However, video marketing is only a valuable addition if you can get your videos found. Video SEO experts take the massive popularity of videos and translate it into content that can be easily found by search engines. Video SEO is an important factor in whether or not your videos will be seen by your target market. With video growing in popularity as a method of content distribution, it's essential that your videos be optimized and easily searchable.

ESSENTIAL COMPONENTS OF EFFECTIVE VIDEO SEO

Your video SEO company should help you distribute your professional marketing videos that will help brand your business, expose your message to a wider audience and get higher search engine rankings. An experienced video SEO company will show you exactly how to best use your video content to meet your marketing goals.

Video SEO is an extension of SEO; so many portions of your evaluation of a video SEO company are similar to evaluating a standard SEO company. Video SEO is a process of "helping search engines understand video content." (Robertson) A video SEO company will help optimize your video and then distribute the video to a network of video sharing sites. They will also promote the videos to increase your traffic, if necessary. In some cases, the video will be hosted on your website as well as 3rd party sites. There are slightly different optimization techniques for each option.

Video content that is hosted on your site will bring traffic to your website and allow your company more control over the monetization of the video. Your company will also be able to have more control over analytics and be able to track exactly how the video is being viewed. On the flip side, 3rd party distributed video content offers a lot more exposure. It is an excellent tool for drawing in a great number viewers in a short period of time, but your viewers have to take an "extra step" in order to get to your site and see your offer.

Although most best practices of SEO come into play, there are also many SEO factors that are specific to videos. Video SEO is about how to frame the video content to get traffic and increase authority for specific keyword terms and for specific parts of the market.

GENERAL RED FLAGS FOR VIDEO SEO

+ **Emphasis on video SEO as a "cure all" for search engine traffic:** As the statistics show, video marketing can definitely increase your search engine traffic. However, video SEO is a small part of your overall Internet marketing strategy. Professional video SEO companies understand this and should emphasize that video SEO should be used in conjunction with other SEO strategies, include landing page optimization. Creating and distributing a video is not a source of instant traffic.

+ **No experience with traditional SEO:** If a video SEO company or consultant brushes off traditional SEO as "a thing of the past" or doesn't show competence in basic qualities of text based SEO (see the SEO chapter of this book), they aren't qualified enough to help your company achieve rankings for your videos. Video SEO shares the same foundational elements as text based SEO.

+ **Guarantees and promises of top placements for videos:** While videos are more likely to rank higher than text content with the same keywords, a properly optimized video is not a guarantee for top placement. A professional video SEO company will explain that their strategies will improve your results, but they shouldn't offer specific guarantees of any type.

+ **Pushing low quality video creation software or packages**: Although some video SEO companies do not handle video creation, some will offer development as part of their SEO services. Watch out for low quality video creation software that is promoted by some video SEO "experts." Normally, these companies are making their real money off of the video creation software and not from their SEO efforts. They may be using low cost SEO services to entice customers to purchase the software, without having any actual SEO experience.

+ **No advice on quality video creation:** Even video SEO companies that don't offer video creation service should have some ideas about what works best and what doesn't. If you have yet to start creating your videos, ask the video SEO company if they have any advice on the type of content you should use. If they can't make any suggestions at all, they probably do not have much experience successfully promoting videos. The quality of the video is an important part of the process, especially if you will be posting it on a social site like YouTube. Your video SEO company should have your best interests in mind and try to help guide you toward making a successful video.

+ **Pressuring you to let them create a video:** With the last point in mind, there is a difference between offering a few tips and pressuring you to scrap your existing video to allow them create one for a higher cost. Professional video SEO companies are interested in making your existing video accessible to the search engines and your market using SEO principles. If video creation is their focus, they may not be experienced enough to help you with SEO. If they do offer video creation services, make sure that their SEO experience is up to par.

GENERAL QUESTIONS ABOUT VIDEO SEO

✓ **How long has your company been working with video SEO?** Since video SEO is a relatively new field, the length of experience may be as little as one to two years. However, by looking at their referrals and client results, you should be able to get a feel for their level of experience and their breadth of knowledge. The longer the track record, the better, but a shorter track record is not necessarily a sign of incompetence.

✓ **What is your experience with text-based SEO?** Look for a company that has a background in traditional text based SEO because the same principles apply to video SEO, albeit with slightly different implementation methods. The SEO chapter of this book includes red flags and questions to ask a text based SEO company that are applicable.

✓ **How can video SEO help my video distribution and my business?** This is a test question to gauge their level of knowledge. Look for an answer that refers to video SEO being a way to help search engines better understand video content so the video content can be properly ranked. They should also emphasize that keyword selection for your video, keyword usage on the hosting page and the layout of the video page are important factors in how the video is seen by search engines and used by visitors.

✓ **Will you create my video for me?** This can be a grey area. Most top video SEO firms are actually SEO firms that offer video promotion and targeted traffic solutions. However, there are a few that come from a video production background. If you need video creation services, it may be in your best interest to search for video creation company AND a video SEO company so you can get the best from both fields. That being said, there are a few companies that do both well. Make sure that they show experience in both SEO and video creation before you hand over your video marketing campaigns to them.

VIDEO KEYWORD ANALYSIS

Selection of keywords for video marketing is very similar to text based SEO. Your video SEO company will likely start aggregating a list of keywords based on your niche and the traffic that your site is already receiving. Since you have an existing site, your incoming traffic will be a good indicator of how your site is already ranking. A professional video SEO company will build upon that incoming traffic to develop a targeted list of keywords for your videos.

If your video has already been produced, your video SEO company should integrate their marketing findings with the content of your video. Slapping on slightly related keywords onto your video is not going to achieve much. The search algorithms for YouTube are getting more precise and simply using a keyword in the title will no longer be enough to get your video ranked. Social aspects, like rating and community labeling, come into play so your video SEO company needs to use keywords that are closely related to your video.

In addition, your video SEO company should be using tools that are specific to video SEO. YouTube released "YouTube Suggest" in mid 2008 which marked the first foray into video specific keyword research tools. (Robertson) As of this writing, it is one of the only tools that is specifically designed to aggregate video based keyword research results, however there are many video search engines (Truveo.com for example) that can help your video SEO company understand the video competition for your market. It's vital that your video SEO company uses video based keyword results, along with traditional keyword methods, because it will produce a more targeted and relevant results for your video. For more on proper keyword research methodologies, see the SEO chapter of this book. Keyword research will create a basis for the video SEO company's strategy for marketing your video, so it's very important that they are knowledgeable in this area.

RED FLAGS FOR VIDEO KEYWORD ANALYSIS

+ **Using keywords without checking their relevance with video content:** A large part of rankings for videos on social sites has to do with how people respond to your video (with comments, ratings, etc). You're less likely to get positive reviews for your video if your content does not relate to your title and your description. In addition, if you are hosting your video on your own site, you're less likely to get back links from others if the video doesn't relate to the keywords.

* **Using keyword research tools that don't relate to video specifically:** Keyword research for video SEO is still undergoing growing pains. The next few years will probably bring even more specific keyword research tools, especially because the popularity of video is rapidly expanding. A professional video SEO company should grow with the industry and utilize the specific tools as they are released to increase your results from video marketing.

QUESTIONS TO ASK ABOUT VIDEO KEYWORD ANALYSIS

✓ **What resources do you use in order to conduct your keyword research?** Look mention of the tools listed previously in order to determine if the company is using industry standard tools.

✓ **How do you account for the difference between keyword results for text based sites vs. keyword results for videos?** A professional video SEO company will explain that text based keyword results will help provide the broad base for the keyword list while more specific research with tools like YouTube Suggest will help tailor those results for videos. If they do not address the difference, they don't have enough experience with working with videos.

ON PAGE OPTIMIZATION

On page optimization for your video includes all of the elements that are used in the tagging, description, title and code of the page where the video is being hosted. As previously mentioned, there are two distinct methods for on page optimization – on page for hosted videos and on page for distributed videos. In this section, we'll be looking at the subtle distinctions between the two methods of posting video content and how your video SEO company can optimize the presentation to draw in more viewers.

Unlike traditional web search, video searchers are approaching content in a different way. This results in video on page optimization practices that utilize attention grabbing, yet relevant, techniques. When individuals perform a web search, they normally type in their keywords and then drill down until they find just what they are looking for. For example, if a searcher is looking for a red shirt, they will likely start by searching for "red shirt" and get more specific with each search until they find "Brooks Brothers red shirt xl mens."

On the other hand, video viewers look at results from a video search engine in a very different way. They will normally type in their keyword term and then scan through the results looking for the most relevant video. They won't refine their search terms again and again. They will use the thumbnail image and the title to determine whether or not your video is worthy of their time.

Even though you may be focused on getting your videos to appear in web searches (like through Google's universal search) understanding the way that videos appear in video search engines is important. Your video needs to appeal to users of video search engines, because their popularity in video search engines will have a profound effect on how your videos show up in organic search engine listings. Furthermore, website searchers are using video sites more than ever to primarily conduct their searches. In December of 2008, YouTube became the second largest search engine on the web. (Robertson) It is highly likely that your target market is already using a video search engine to find information on your topic.

In addition to needing to meet the requirements of two different types of searchers, your video SEO company will also need to meet the requirements of two different types of search engines. There are first generation and second generation search engines that deliver video results. First generation video search engines are traditional search engines that have adapted their search engine algorithms to search for video content specifically. For example, Google Video search (http://video.google.com) and Yahoo Video search (http://www.yahoo.com/ - click "video" above the search box) both apply their existing algorithms to video content in order to rank them. First generation video search engines are primarily concerned with the META data of the video and the on-page text.

Second generation video search engines (Truveo.com, Blinkx.com and Veoh are a few examples) approach indexing videos slightly differently than first generation search engines. They utilize META data and on page text, but also incorporate user generated content (comments, tags and annotations) into their ranking of videos. Your video SEO company should be able to use traditional SEO methods in combination with methods that will attract positive user interaction.

The following elements should be used by your video SEO company for on page optimization, no matter where the video is being hosted.

o   *Title usage*

Your title is your video's introduction to your market. It needs to incorporate important keywords as well as being entertaining. Your video SEO company should come up with several title options that include a keyword yet have enough attention grabbing elements to make your video stand out. The keyword will attract search engines and the nature of the title will be appealing to human visitors. If your video is hosted on your own site, the title is still important because most people will use your title when linking back to your site. With a strong title, you can be sure that your video will get more clicks on video sharing sites as well as more traffic from other sites where people are linking to your video.

o   *Video File Name*

Titles are viewable by web viewers, but they aren't the only important factor in labeling your video content. The video file name is also important for SEO purposes because the search engine spiders use the name of the file to classify it. "Myvideo. flv" is not descriptive enough to attract attention from the search engines. Video file names that include keywords or the descriptive SEO title of the video will help the search engines be able to correctly label the video, which can increase search engine results.

o   *Length*

The length of your video is important from an SEO perspective. Your video is more likely to be viewed if it is kept short - at just a few minutes. People are busy online and they don't have time to watch an hour or even a 20 minute presentation. Your video SEO company should advise you that bite sized chunks of online video will be more effective and will garner more popularity (and therefore higher search engine results).

o   *Creating a positive user experience*

Just like with written content, the quality of your video content will have a big effect on how well it is ranked. To become an authority in your niche, you have to become a relied upon expert. Since viewers are allowed to rate, comment and otherwise tag your videos it behooves your company to create valuable content. Your video SEO company will be much more successful in garnering higher search engine rankings and traffic if the campaign is based on a solid foundation of a quality video.

ON PAGE OPTIMIZATION FOR HOSTED VIDEOS

Hosted videos are hosted on your website and allow your company to become a trusted resource for information on your target topic. Hosted videos encourage backlinks to your main website, which can increase your rankings overall. When your video is placed on your website rather than being distributed, your video SEO company has a bit more control over how the video is optimized.

As stated in the previous section, user experience, video file name, length of video and the title of the video are all very important to improving SEO for your hosted video. There are also many other factors that your video SEO company should address when helping you place a video on your website.

The first aspect is the design of the video page. Your video SEO company should create a separate URL for your video that includes the optimized video title. Giving your video its own page will help that page rank for the specific keyword terms that your video SEO company has selected.

The page URL can be optimized for the keyword term, but it should also be placed in a smart file structure. For example, http://www.example.com/video/tips-for-saving-money.html is preferable to http://www.example.com/stuffilike/mytips.html. The first example not only identifies the page as a video file but it clearly describes what the video is referring to. Your video SEO company should utilize the page URL structure to clearly label your video.

In addition to the URL, giving your video its own separate page also allows for the use of traditional SEO techniques. The META tags and header information can be optimized for keyword terms relevant to the video. Header tags can be used on the page to call attention to important keywords. The video should be imbedded in the page and not delivered as a java script pop up. Not only is a pop up distracting to the user, but embedding the file will give your video SEO company another opportunity to use your keywords.

META data on the page is important, but many video formats also allow META data in the video code as well. The benefit of using META data in the code of the video is twofold. Many search engines scan the video code for additional ways to categorize the video information, so using keywords in this area can help boost keyword relevancy. In addition, the encoded META data will travel with the video if it is imbedded on another website. When other individuals use your video on their site, your keywords and website in the encoded META data will accompany the video file and increase your back links and keyword relevancy. "In file" META data is only viewable for hosted video files and will not work if you are uploading the video to a video sharing site.

The video landing page should include relevant and keyword optimized content. A description of the content of the video using keyword terms will enhance visitor experience as well as optimize the page. The content will inform the user on what the video is about and will also provide your video SEO company another opportunity for using your keywords.

Finally, your video SEO company should be capable of determining your video's loading time and taking steps to improve loading speed. Users will click away if the video takes too long to load. A professional video SEO company should suggest methods for improving load time, including upgrading your hosting package, and should test the video load time across many different types of connections.

Not only will visitors click away from your video, but search engines will be less likely to display your video as a relevant result because it is taking so long to load. Search engines want to direct traffic to locations that meet the needs of searchers. If your video is loading slowly or not at all, it does not represent a quality result.

Your video is not displayed in a vacuum. Just like other types of pages in your website, the video display page can be designed in such a way to create a positive user experience as well as an optimized structure. Your video SEO company should have the resources to design a video landing page for you, or a common template that they use in order to create a video section on websites.

Dynamically created video pages will increase the visitor's time on your site and increase the likelihood of backlinks from other sources. "Value added" elements on your video landing page can include related photos, audio recordings, news, links to blog posts and links to other videos. The additional links and text on the page will give your video SEO company additional opportunities to tell the search engines what your video is about.

In this example from Fox Sports News (powered by MSN), the main video is prominently displayed with a descriptive sentence along with other related videos that are displayed (with thumbnail photos of their content). If you were to look at the HTML code for this page, you'd see keyword usage in the META tags and headers for the page, which adds extra optimization for first generation video search engines. Your video page does not have to include links to other videos. If your hosted video is the only video on your website, your video SEO company can list additional site pages, blog posts or other related, and helpful, content in order to increase the experience for your user.

However, one element that this particular page is missing is user interaction, which is helpful for ranking in second generation search engine sites. Allowing user comments, adding a rating scale and allowing users to share videos would increase the ranking of the videos on this site in second generation search engines. Giving your visitors the ability to share your videos (by imbedding them in their own sites) will greatly increase your inbound links. Commenting provides additional text on your video page.

For example, if you have a video on pruning rose bushes and one of your visitors comments that they use the same technique with camellia bushes, the phrase "camellia bush pruning" will become associated with the video. Since you didn't mention camellias in the description of your video, this provides an additional channel of traffic that you may not have thought of before. Your video SEO company should implement these community based tools on your hosted video page in order to increase your page's rankings.

ON PAGE OPTIMIZATION FOR DISTRIBUTED VIDEOS

Distributed videos are what most business owners think of when they think of using video to promote their business. Although hosted videos can increase traffic to your main site, distributed videos are much more likely to be viewed by members of your target market. Submitting a video to YouTube, Blip or MetaCafe (all popular video sharing sites) can create a lot of traffic and exposure for your business.

Just like with second generation video search engines, video sharing sites rank the videos based on the number of views, ratings of the video and how much the video has been shared with others. These community based aspects are just as important as keyword usage when it comes to optimizing videos for this distribution channel. The following examples deal exclusively with Google, but the options are very similar for other video uploading sites. In addition, since YouTube is the most popular and largest video sharing site it makes sense to look at their featured and policies in order to determine the best SEO practices for uploaded videos.

According to Mark Robertson of ReelSeo.com, there are several factors that determine ranking in YouTube which can be applied to other video sharing sites as well.

o   Title

o   Description

o   Tags

- o Views

- o Ratings

- o Playlist additions – how many individual users have categorized your video in their personal playlists

- o Flagging

- o Sharing

- o Commenting

- o The age of the video

- o The views on your overall channel

- o The number of users subscribed to your channel

- o How many times the video has been embedded

- o The number of inbound links

(Robertson)

A professional video SEO company has many tools at their disposal to maximize these factors and get your video higher rankings. Proper usage of keywords in the title, description and tags are essential. However, there are many community based factors that your video SEO company needs to deal with. Their first goal should be to attract attention for your video through an interesting title, short description and quality thumbnail image.

When an individual searches on YouTube, the title, description and thumbnail image are all they have to work with to determine whether or not they want to view your video. Since so many factors for ranking are dependent on your video being viewed (comments, sharing, inbound links, etc) it makes sense for your video SEO company to try to maximize viewership.

The title should be enticing but also include important key phrases. There's a delicate balance between creating a title that ranks well for a keyword term and one that will pique the interest of searchers. A professional video SEO company should be able to develop a title that meets both needs.

The descriptive text also allows your video SEO company to utilize keyword phrases. There are two forms of descriptive text used in distributed videos. There are short descriptions of a sentence long that are displayed in the video search results. There is also a large text box that is displayed when the video is viewed. Both of these descriptions should be optimized for keywords, but only the short description will be responsible for attracting attention from searchers.

(source: http://www.youtube.com/results?search_query=super+bowl+2010&search_type=&aq=f)

In this image, you can also see the importance of thumbnail images in attracting attention for your video. Your video SEO company should select an image that is representative of your video but also interesting to the user. The thumbnail image is very important to encouraging visitor clicks.

In the related videos section of video sharing sites, only the title and the thumbnail image will be used to attract the attention of the user. This makes it even clearer that the thumbnail image is important in garnering those all important clicks.

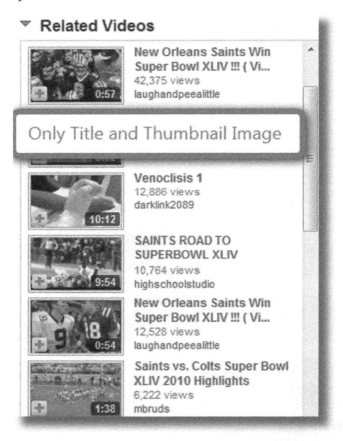

(source: http://www.youtube.com/watch?v=4w6VLkSiG2Q)

Your video SEO company should be familiar with creating a compelling thumbnail image and apply that experience to your video upload. Along with the thumbnail image and the short descriptive text, your video SEO company will be responsible for making use of the long descriptive text which will be displayed along with your video.

(source: http://www.youtube.com/watch?v=4w6VLkSiG2Q)

Your video SEO company should add your URL to the long description within the first few sentences. The description can also include keyword rich text which is mirrored in the tags that are used to describe the video. Making good use of these areas when a video is uploaded will greatly increase your videos keyword relevancy and will help your video get more views.

YouTube and other video sharing sites are free to upload, but there are also paid options that may get your video more exposure. Paid placement will put your video in the top placement for related videos and also at the top of search results for related keywords. In addition, "featured videos" as they are called can be featured on the homepage of YouTube for specific users who have searched for related content in the past.

Annotations are another tool that your video SEO company can use in order to increase your video views and get your website more traffic. Annotations come in three different varieties – speech bubbles, text blocks and marquee areas. They will appear over the video content and can be used to add information to the video.

More importantly, from an SEO perspective, your video SEO company can use them to link two videos together. This means that while users are viewing your video they can instantly click on the annotation to see your additional videos.

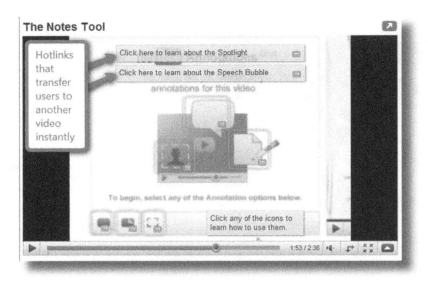

(source: http://www.youtube.com/watch?v=lxQ1b8KR-Qo&annotation_id=annotation_7
01885&feature=iv)

By utilizing annotations, your video SEO company can create cohesion between the videos they have uploaded for you which can build more interest in your brand, and in your channel. Your "channel" comprises all of the videos that your company has uploaded. The channel page gives your company yet another platform for optimizing your company's online video presence. When your video SEO company creates a YouTube account, they will be able to create a detailed description of your site that can include key phrases and a compelling description.

In addition to this channel description, the channel page will display your most recent video at the top with a list of your most recent videos underneath. This channel profile leads us to the next important aspect of video SEO for hosted sites – community. Professional video SEO companies will also make use of this feature by having your user account subscribe to other user accounts with similar topics. Getting involved in the community of YouTube in a strategic way will help the optimization of your channel and your videos.

Each video has various social aspects that allow users to rate and give feedback on the videos. Your video SEO company should be rating relevant videos, making connections with other users in the same channel and adding responses when it is appropriate. Your videos can even be used as comments to other videos, kind of like a video reply. Your reply video will be seen underneath the primary video, which can get more views for your video.

Video responses are displayed in a special featured area of the video screen, which will increase their likelihood of being clicked and viewed. By strategically using video responses, your video SEO company will be able to build off of the success of related videos and increase the SEO power of your video submissions.

VIDEO TRANSCRIPTION AND CLOSED CAPTIONING

A final tool that your video SEO company should use to optimize hosted or distributed videos is video transcription. Mentioned earlier, video transcription will allow the search engines to understand what your video is about because the content will be translated into readable text. Closed captioning is another option that allows your video SEO company the opportunity to provide more relatable content to search engines.

Closed captions enhance user experience, which is good for SEO, but they also give the search engines something that they can work with in order to index and classify the content of the video. Google search engine spiders read closed caption files as well as external transcripts. Yahoo will only read text transcripts that are attached to the file or present on the page. Blinx. com, a popular video search engine, will actually do speech recognition on the video in order to determine the content of the video and classify it properly. (Robertson)

Your video SEO company should use one of the available methods for captioning. They are:

o   Open captions – These will be synchronized with the video and are viewable to all viewers of the video. The major drawback for these is that the captions cannot be turned off, which can be distracting to some viewers.

255

o Closed captions – These are imbedded as part of the video file and can be opted into by viewers who need closed captioning services.

o Link – This is present on the video landing page and will transport visitors to a text transcript of the video.

o On page display – This form displays the transcript of the video just below the video display so users, and search engines, can easily read the content.

o External subtitle file – This uses a timed text file or other type of subtitle file.

Your video SEO company should use one of these forms of transcription in order to increase the keyword relevancy for your video. Just as a note, YouTube and some other video sharing sites offer speech-to-text capabilities on their videos, but it is not always accurate. Similarly, if your video SEO company will be utilizing speech-to-text software, it should be double checked by a human reader before it is posted on your site.

RED FLAGS FOR ON PAGE OPTIMIZATION

+ **Focus on either hosted videos or distributed videos as superior:** Each method of video distribution is equally as important. They both have specific benefits that companies such as yours can benefit from. If a video SEO company is too focused on one method of distribution, they won't be effective in helping you create a multi-dimensional video marketing plan.

+ **Focused on obtaining rankings in Google alone:** Although Google is still the most popular search engine it is only a small piece of the video marketing puzzle. Your video SEO company should be capable of optimizing your video so that it will rank well in Google, on second generation video search engines and within video sharing sites as well. These three platforms represent the ways that users find videos, and a professional

SEO company should be able to create an audience from each area.

+ **Poor title creation:** By looking at examples of previous videos optimized by an SEO company, you can determine how well the video SEO company is able to create video titles. Considering the fact that the title of your video is a linchpin in your video's popularity, and therefore it's ranking, crafting titles that are appealing and that use keywords is an essential part of video SEO success.

+ **Poor thumbnail creation:** Just like the titles of the videos, the thumbnails are essential in getting all important clicks to your videos. Thumbnails are very effective in increasing viewership in video sharing sites and video search engines, which translates to better SEO. Look at examples of previously created thumbnails that your potential SEO company has done for other clients. The thumbnails should be clear images that directly show what the content of a video is all about.

+ **Poor short and long description creation:** Again, look at examples of how the video SEO company is utilizing short and long descriptions with other videos. Look for succinct explanations of the video content that is not only informative but appealing to the target market.

+ **Focused on optimization and not enhancing user experience:** User response to a video is becoming more and more of a factor in whether or not a video ranks highly in search engine results, in both types of search engines. Keyword optimization is important, but it is not the be all and end all of video SEO. Your video SEO company should be familiar with the previously outlined techniques for enhancing user experience and assist your company with taking steps necessary to make the viewer's experience a quality one.

+ **No experience in designing a video landing page:** Your video SEO company should be familiar with the basic methods outlined above for designing a video landing page and making it appealing for the viewers. If they don't have a web design team in house, they should refer one to you or they should work with your existing designer.

+ **Insisting on submitting to a second tier video sharing site rather than YouTube:** If submitting videos is part of your video marketing strategy, YouTube should be a primary focus for your video SEO company. Other second tier video sharing sites are helpful in enhancing your efforts, but they shouldn't be used in place of YouTube. YouTube should be the primary target for video distribution. If a video SEO company is trying to convince you that submitting only to a second tier site will be in your best interest, they are trying to inflate their results. Since second tier sites may have less competition, your video SEO company will get higher rankings within the system immediately, which can artificially improve their SEO results.

+ **No help with channel development for YouTube:** Your channel landing page is important for growing keyword relevancy for all of your videos as a whole, as well as enhancing the community interaction aspect of video sharing sites.

+ **No interaction with other video sharing site users:** Your video SEO company should be familiar with various methods for interacting with the video sharing site community. Interaction helps ranking and relevancy so this is an area that should not be ignored.

+ **Lack of knowledge about the importance of imbedded content:** Although imbedded content and transcripts are relatively new phenomenon, an experienced video SEO company should be ahead of the curve when it comes to speech-to-text technology and the other forms of imbedding text within video content. They should have resources for translating the spoken content into text that is readable by users and search engines alike.

QUESTIONS TO ASK ABOUT ON PAGE OPTIMIZATION

✓ **How do you approach first generation search engine optimization vs. second generation search engine optimization?** Your video SEO company should be familiar with the difference between trying to rank in Google and trying to rank in Veoh. They should be able to tell you that while Google needs to be courted with keywords and backlinks, Veoh depends on those factors *plus* social elements. Your video SEO company needs a firm grasp of how to rank well in both of these platforms.

✓ **How will the keywords you discovered be used in my video distribution?** Your video SEO company should explain that the keywords will be used in the title, tags, short description, long description, META tags and other on page elements. Make sure they have a comprehensive understanding of keyword usage.

✓ **How will my video page encourage people to participate?** If hosting the video is part of your strategy, your video SEO company should be able to identify and utilize many on page elements to encourage user interaction. Video comments, embed links and ratings scales will help your visitors interact with your site and improve the traffic (and SEO) of your video page.

✓ **What should I do about slow videos?** Your video SEO company should encourage you to speak with your hosting company about adding more bandwidth to your account, or they can advise you on compressing the size of the file. These steps will help the video load faster, no matter what type of connection your visitors are using. This will greatly improve the user experience and improve rankings.

✓ **Will you help design a video landing page or do you have a custom one you use?** Video landing pages for hosted sites is essential in making those pages effective. Your video SEO company should be well versed in how to design a video landing page. Their experience in this area should come from working with a web design firm or from someone in house. Ask to see examples of their standard video landing pages so that you can assess the quality.

✓ **How will you encourage participation on YouTube?** Distributing a video on YouTube and increasing its rankings in their search engine will require social interaction. Your video SEO company should advise using your videos as video responses to other related videos, strategically making comments on other videos and subscribing to related channels as tools for increasing participation from other related users.

✓ **How are annotations used on YouTube and will you be using them?** Annotations can really help drive traffic to your other videos, so your video SEO company should look for opportunities to use them. They should be very familiar with the different annotation forms and how to use them effectively to enhance the quality of your videos and link them together.

✓ **Will you use closed captioning files or transcripts to add text to my videos?** The answer to this question should be yes. An experienced video SEO company will know the value of having text transcripts of the video content for SEO purposes. No matter what the method is, your video SEO company should utilize a human reader to confirm that the text matches the video.

OFF PAGE OPTIMIZATION

Off page optimization involves the links that are coming from other websites to your website. The more backlinks your video has, the more it will be seen by the search engines as an authority site, and your search engine rankings will increase. Your video SEO company can encourage visitors to create a backlink to your site through various methods and can also create backlinks to your website themselves. Following are a few key factors and tools that your video SEO company should be using in order to create a strong backlink profile for your video.

QUALITY MATTERS

The biggest factor in whether or not users will voluntarily create backlinks to your website is the quality of your video. Since this is primarily under your control, you should take every step necessary to create a quality video. Better production values, clarified content and helpful information will outshine other videos in your market that were put together hastily. Your video SEO company can only do so much with a video. Even the best company cannot coerce people to create backlinks to your video landing page. Assist them with helping you by creating a quality video to begin with.

ALLOW USERS TO EMBED YOUR VIDEOS

Although traditional one way links (see the linkbuilding chapter of this book) are helpful from an SEO perspective, video files have an extra method of distribution. Videos can be embedded in other websites and viewed from those websites. When a visitor embeds your video in their website, all of the META data goes with it along with a link back to your site.

Your video SEO company should assist you with creating an embed code area on your video landing page to make it easy for your visitors to utilize your video on their site. Within just a few lines of code, they'll get helpful content that they can use on their site and you'll get a quality backlink. YouTube and other video sharing sites automatically aggregate embed codes so that your video can be used wherever the visitor deems fit. Be sure that your video SEO company uses an HTML wrapper as part of the embed code because this will ensure that the video backlink will be using the correct anchor text.

CROSS PROMOTIONAL LINKS

There are two distinct ways that your video SEO company can use cross promotional linking to improve the amount of backlinks to your video landing page, or your hosted video page. First, if your company has several web properties, your video SEO company can embed your video on each of your websites which creates relevant links back to your video page. This strategy will work as long as the other content on your web property is related to your video's content.

Another form of cross-promotional linking is creating link trades with other websites that have similar content. Your video SEO company should work to establish relationships with related companies that may be interested in embedding your video on their site. Related blogs are another good source for cross-promotional linking. Blogs are visited more regularly than standard websites and are more likely to have a group of followers that will respond to your video.

SOCIAL MEDIA OFFSITE OPTIMIZATION

You Tube and other video sharing sites automatically make it easy for your video to be shared among the various social networks. Your video SEO company should also make it easy for visitors to add your video to social networking and social bookmarking sites if your video is hosted on your own website. Sharing icons, which are displayed on the page next to your video, will allow visitors to spread the word about your video with just one click. Digg, StumbleUpon, Facebook, Delicious, Reddit and Twitter are all excellent sources for backlinks and can be easily utilized with sharing icons. For more on effective backlink building and social media strategies for your video landing page, see the link building and social media optimization sections of this book.

RED FLAGS FOR OFFSITE OPTIMIZATION

- **Ignoring offsite optimization methods:** Like any standard web page, your video landing page should be receiving offsite backlinks in order to increase its rankings. Although videos are getting more traffic, these pages still need backlinks. Even a video posted on a video sharing site can benefit from the use of offsite optimization strategies.

- **Not adding embed codes to a hosted video page:** Inexperienced video SEO companies aren't familiar with the backlinks that can be created through embed codes. Although it would see to the untrained eye that this practice is giving content away for free, the actuality is that the HTML wrapper in the embed code is boosting the optimization for the hosted video landing page.

✦ **Not encouraging social media:** When it comes to videos, social media is a powerful ally. People are more likely to share videos than any other form of media. Your video SEO company should make it easy for users to share and comment on your videos.

QUESTIONS TO ASK ABOUT OFFSITE OPTIMIZATION

✓ **How will your company help increase the amount of backlinks for my hosted video page?** Your potential video SEO company should describe the various methods of obtaining backlinks, including encouraging social bookmarking, allowing embedding of video files and making connections with webmasters and bloggers in related fields.

✓ **How do you encourage users to embed and distribute my video content?** If the video content is hosted on your website, the video SEO company should make use of embed codes so others can host the video on their sites.

REACH

A professional video SEO company will be able to assist your company with submitting your video to a wide variety of sources in order to increase your visibility, brand your business and drive traffic back to your site. Their reach will be determined by how many sites and platforms they can use to publish and promote your video.

Generally speaking, your video SEO company should at least be familiar with YouTube and hosting videos on your own website. However, there are many other video sharing sites that should be considered. Look for a video SEO company that has considerable with several of the following.

o Yahoo Video – http://video.yahoo.com

o Video.qq – http://video.qq.com

o You.video.sina.dn – http://v.sina.com.cn

o   Flickr – http://www.flickr.com

o   Photobucket – http://photobucket.com/recent/videos

o   Youku – http://www.youku.com

o   Daily Motion – http://www.dailymotion.com

o   Rediff – http://is.rediff.com

o   Rambler Vision – http://vision.rambler.ru

o   4Shared – http://www.4shared.com

These sites represent the top 10 in terms of Alexa rating, but there is a complete list of all known video sharing sites at http://www.reelseo.com/list-video-sharing-websites/. Your video SEO company should be experienced with as many sites as possible and have a complete promotional plan for your video distribution to get the most impact.

It's essential for your video SEO company to be experienced with a variety of video sharing sites because each has it's own rules for submission. Although the sites serve the same purpose, there may be distinctions between preferred optimization methods that only an experienced video SEO company will be able to utilize effectively.

SUBMISSION THROUGH FEEDS

Your video SEO company should utilize video feeds in order to let search engines know about the existence of your video. Although videos will eventually be indexed by search engines, it's important that your video SEO company increase your reach by submitting RSS feeds. There are two significant types of RSS feeds for video – RSS and mRSS. They can both be essential in getting traffic for your video.

A professional video SEO company will be able to create and use both types of feeds to improve reach. An RSS feed will allow visitors to subscribe to your content and your updates when your video is hosted on your website. They will help improve search engine rankings in traditional search engines, like Google's universal search.

mRSS feeds are designed for video search engines. They allow your video SEO company to digitally inform second generation search engines about the existence of your video. mRSS feeds will contain the META data, URL of the content, thumbnails, video ratings, play counts and tags. For this reason, the mRSS feeds should be aggregated after the video has been present on a video sharing site for some time.

Your video SEO company should also create a video site map. While these aren't as effective as RSS feeds, they can only help increase the traffic and rankings for your video. They are important because Google does not accept mRSS feeds, but they do accept video sitemaps. Since Google is so important to SEO in general, it makes sense that your video SEO should take the time to use a video site map.

RED FLAGS FOR REACH

+ **Focusing attention to submitting to one video search engine:** After your video SEO company hits the "big guns" of YouTube, Google video and Yahoo video, they should move on to other video submission sites to get your video the widest reach possible.

+ **Ignoring feeds and video site maps:** These are powerful traffic tools that professional video SEO companies should be very familiar with. There is no excuse for leaving these tools out of the process.

QUESTIONS TO ASK ABOUT REACH

✓ **Where do you regularly submit client videos?** Look for a response that indicates that they regularly use a variety of different video submission sites. The more experience they have with submitting on different platforms, the better reach they will have.

✓ **What is the difference between mRSS feeds and video site maps?** Your video SEO company should respond that mRSS feeds are used for all other search engines, besides Google which accepts video site maps. Both are essential for effective reach.

REPORTING METHODS

Your video SEO company should have a specific procedure for analyzing and reporting the results from your video marketing efforts. Using analytics software installed on your hosting will let the video SEO company know how many people are visiting your hosted video and viewing the video. They will also be able to track the comments, interactions and embeds from your hosted video.

Each video sharing site has its own analytics data that your video SEO company will have to contend with. When it comes to video SEO many of the analytics procedures are still being defined and refined, so it will require experience from your video SEO company to correctly identify how well your videos are performing and what type of results are being garnered from video marketing.

Although tracking video SEO results is somewhat complicated, it's essential that your company has a reliable method for determining results. Without these determinations you will not be able to measure your results from video marketing. An experienced video SEO company should be able to track data and give your company suggestions for further optimization.

RED FLAGS FOR REPORTING METHODS

+ **No clear plan for tracking and reporting:** Your video SEO company should be able to compile the data and deliver it to you in an organized format. A company that expects you to do the tracking and analytics work is inexperienced and a waste of your time.

+ **No site specific video tracking methods:** The video SEO company should be familiar with tracking data from each type of video sharing site so they can create accurate reports.

QUESTIONS TO ASK ABOUT REPORTING METHODS

✓ **How will you deliver the results of your video SEO efforts to me? How often?** Understanding this at the onset of your relationship with your SEO company will help define your expectations of their service.

✓ **Can I see an example of a video SEO report from a previous client?** Even if your video SEO company describes their reporting methods, it can be difficult to visualize how this data will be delivered to you each month. Seeing an example can help you assess whether or not their reporting methods will be helpful to you.

CONCLUSION

Video SEO is going to become more important in the coming years and there may be many SEO experts entering the field that may or may not have enough experience with video SEO. Be sure to do your due diligence with this type of service provider to ensure that your videos will be in good hands. Video SEO is a specific type of SEO and should not be placed in the hands of an inexperienced provider or even someone who is experienced in traditional SEO. The more experience that a company can show you that directly relates to video SEO, the better they will be at meeting their needs.

CHAPTER NINE:

LOCAL SEO FIRMS

Local search engine optimization takes the best practices of SEO and combines them with techniques that will get your company noticed in your local area. With more and more people turning to the Internet rather than the phone book when they are trying to find local services, time spent with a local SEO specialist can pay off in the form of more local customers.

A local SEO company will evaluate your website and determine what specific steps need to be taken to optimize your site for locally related keywords. In addition to helping your site rank well for local search in Google they will also attempt to increase your rankings in industry and location based search engines.

The following chapter covers the best practices of local SEO and the major areas you should look for when working with a local SEO company. To enhance your understanding of some of the concepts in this chapter, please refer to the SEO chapter of this book which goes into greater detail about search engine optimization. In addition to the following criteria, you can see the latest listings for local search engine optimization companies at http://www.topseos.com/rankings-of-best-local-search-companies.

THE IMPORTANCE OF LOCAL SEO

Local SEO has been growing in importance in the last several years and is absolutely essential if you sell products to local customers. Local customers are searching for your product or service online and with the advent of smart phones more people are looking for products and services while they are on the go. You have to be found online when people look for terms related to your business.

It's advantageous to you to target these locally searching customers. When people start searching for product with local phrases added, they are further along in the buying cycle. They are more likely to be closer to making a purchase and they are more qualified. For example, if someone is looking up the word "landscaping" they might just be looking for general landscaping advice. However, if they are looking for "landscaping Providence area" they are looking for landscaping help in that specific area and are closer to making a purchasing decision.

Local SEO is not only about getting your main website to rank for your keyword terms, but it also involves getting your traffic to the right pages of your website. For example, if someone in Los Angeles, California is looking for a chiropractor, you can direct the searcher to the page of your website that includes your address, phone number and a map. This will put searchers right where they want to be – especially if they are using a mobile device to locate information in their local area. Directing them to a page that includes contact information rather than a helpful article will meet the needs of the local searcher.

Even if the searcher doesn't type in a geographical location, Google, Yahoo and Bing will normally deliver local results at the top of the search engine page for specific types of searches. The search engine will pinpoint their location from their IP address and will deliver results based on that location.

Google, in particular, has made local search engine results a priority. Local business results are delivered in a grouping at the top of search engine results page, along with a map that identifies the locations of the business. It's often referred to as the "Google 10 pack" because up to 10 local businesses are displayed from the search. Here's a 10 pack example from a search for "pet shop" from an IP address located near Fowler, California. The 10 pack was listed under 3 general search results related to "pet shop."

Local business results for **pet shop** near Fowler, CA - Change location

A — Whitie's Pet Shop, Inc., dba Whitie's Pets
www.whitiespets.com - (559) 438-4343 -
12 reviews

B — PetSmart Clovis
stores.petsmart.com - (559) 297-9514 -
14 reviews

C — Pet Extreme (Clovis)
www.petextreme.com - (559) 298-9738 -
6 reviews

D — SPCA Animal Shelter
maps.google.com - (559) 233-7722 -
5 reviews

E — Aquarius Aquarium Inc
www.aquariusaquarium.com -
(559) 224-3474 - More

F — Seven Seas Pet Shop
maps.google.com - (559) 298-4091 -
2 reviews

G — Aquatic Pets & A-Plus Ponds
www.aquaticpets.net - (559) 298-4549 -
4 reviews

More results near Fowler, CA »

(source: http://www.google.com/#hl=en&safe=active&q=pet+shop&aq=f&aqi=g7g-s1g2&aql=&oq=&gs_rfai=&fp=ae8f9588018abe0f)

If you own a local brick and mortar or service business, it's absolutely essential that you optimize your page for your local area. A qualified local SEO company will be able to use the techniques of search engine optimization and apply them to your local setting, including accounting for specialized local search engine results (such as the Google "10 pack").

NEEDS ANALYSIS

Just as with standard SEO practices, needs analysis is the essential foundation of good local SEO practice. A qualified local search company will begin by analyzing your site's current rankings for your business type and your local area. They will determine how each page is ranking, where the majority of the traffic is coming from and what needs to be accomplished in order to better optimize the site.

They should also ask you about your goals for your website. What type of customers are you trying to bring in from your local area? Where does your company typically do business and where would you like to expand your business? Does your company have more than one location, and does each location have a separate URL? The answers to these questions, and more specific questions, will help a skilled local SEO company determine the best course of action for your local website.

Finally, your local SEO company will likely look at your competition in order to determine how they are ranking for local keywords and what optimization techniques they are using. This will help them determine the lay of the land for your niche and for your area.

RED FLAGS FOR NEEDS ANALYSIS

+ **Lack of customized needs analysis:** Without understanding your needs, a local SEO company can't possibly make an impact on your rankings unless they understand how your site is currently performing. Blanket changes to your website aren't very effective and aren't how professional local SEO companies operate.

+ **Not asking about your major competitors:** If you've been in business for any length of time, you know who your competitors are and will be able to inform your local SEO company about them. The local SEO company should ask about them and should make this information a part of their initial consultation with you.

QUESTIONS TO ASK ABOUT NEEDS ANALYSIS

✓ **How will your company analyze my current website rankings for my keywords?** Your local SEO company should explain their process of needs analysis and define *how they'll do research on your site.*

✓ **Will you be encouraging the right type of visitors and discouraging the wrong type of visitors?** A skilled local SEO company should be able to use selective keyword processes to encourage targeted traffic and discourage traffic that is just looking for general information on the topic.

KEYWORD SELECTION AND LOCALIZATION

Keyword research is essential for any type of search engine optimization; however, the keyword research process for local SEO is slightly different. It's much more than just putting a local keyword in front of your main keyword. In order to get the most benefit from local SEO, each location needs to be analyzed individually. If your business extends to several different cities within your geographic location, your local SEO company should be analyzing each area as a separate entity.

In addition, the terms for each area may be different depending on the preferences of that local area, called localization. While people in Phoenix may look for "garden landscapers in Phoenix" people in Scottsdale might be looking for "Scottsdale landscaping." Your local SEO company should be aware of these differences and test for them. Even though this level of testing can be tedious, a professional local SEO company should take the time to do it. Don't allow a company to skimp on this area.

Another area of localization that your local SEO company should be able to detect is the use of neighborhood names rather than city names. Neighborhood names are just as localized as official city names, so you should make your SEO company aware of any major neighborhoods in your area that deserve their own keywords. This can make a big difference in getting your company's website in front of the right group of people.

If you have several locations, take care to make sure that each location has its own URL and that the content on each site is slightly different. Even though your separate locations may offer the same exact service, your local SEO company should direct you to create unique content for each location specific site, especially if they are being hosted on the same URL. For example, if you have a page optimized for "Phoenix landscaper" and a page optimized for "Scottsdale landscaper" they should include slightly different content even though they are basically on the same subject. This will help your site avoid being penalized for duplicate content.

RED FLAGS FOR KEYWORD SELECTION AND LOCALIZATION

+ **Applying universal keyword search terms to local search:**
  Local search optimization requires more in depth keyword analysis that relates directly to your local area. An experienced SEO firm without local experience will not take the extra step

273

toward making sure that the local terms are relevant to your specific local area.

+ **Spinning the same content for each page with different keywords:** Reusing content on several pages on your website with a different localized keyword is setting your site up for problems with the search engines.

QUESTIONS TO ASK ABOUT KEYWORD SELECTION

✓ **How will you determine whether or not my keywords are targeted for my specific local area?** Your local SEO company should do extensive research on your local area, including asking you questions about your local area if they are not in the same geographic area. They should do research on your local competitors and make a list of relevant keywords based on their keywords.

✓ **How will you address different geographical areas that my company deals with on my website?** A common solution is to develop several URLs for each location (i.e.: Phoenixlandscaper. com, Scottsdalelandscaper.com, etc. If this isn't possible, your local SEO company should work to create unique areas on your site that are targeted for each geographic area.

OPTIMIZATION

Once your local SEO company has selected your ideal keywords, they will utilize on page and off page optimization techniques to ensure that your site will rank well for those keywords.

ON PAGE OPTIMIZATION

As discussed in the general SEO section of this book, on page optimization techniques such as the proper use of META tags will help optimize a page for a particular keyword. In addition to optimizing META tags, your local SEO company should also make use of fresh content on your website that uses local keywords.

By adding several informational content pages to your site, they'll be able to create relevancy between your site and your keyword terms. For example, hosting an article titled "10 Ways to Find a New Home in Phoenix" that uses the term "Phoenix real estate" frequently will add relevance between your website and that keyword phrase.

Another optimization technique should include putting the business address and phone number on all of the pages of the site. This will result in your address and phone number showing up as related results for your keyword terms and increase the authority of your site. Keep your address, phone number and business name consistent across the breadth of your site and across the directories. If you don't have a standard way of expressing your address (i.e.: Ave. instead of Avenue) now is the time to develop a standard so that everything is uniform.

The company should also translate your vital location information into hCard format. hCard format is a simple format that represents people, companies, organizations and places. hCard format will add another touch point for search engine spiders that links your company's website to your location and will increase your chances for appearing in the search engine results for your area's relevant keywords.

Your local SEO company should include other areas that you serve as part of your META tags and in your website text. For example "Sun Real Estate – meeting real estate needs in Phoenix, Scottsdale, Tempe and Guadalupe" as text footer on each page will increase relevancy for those locations and place your site higher in the search engine results for related terms.

Most web development companies will be familiar with XML site maps, and KML files are similar in function. A KML file is a location map and should be utilized by your local SEO company to link the longitude and latitude coordinates of your business with your website. It gives your URL an addition link to your physical location and gives search engines more confidence in using your website as an appropriate result for your location keyword terms.

OFF PAGE OPTIMIZATION

Off page optimization techniques include those discussed in the SEO section of this book, but there are also several other methods that are appropriate for local optimization. Once your site has been optimized on page, your local SEO company should submit your URL to several different directories that rank and list local services.

These third party directories will provide your company an additional page online where your contact information and URL is clearly displayed. There are several categories of directories that your local SEO company should use. When submitting to these sites, the company should use the selected keywords to increase your optimization for those terms.

MAJOR SEARCH ENGINES WITH LOCAL SEARCH CAPABILITIES

o   Google Maps (http://www.google.com/local) – Google Maps is the most frequently used local directory. There is a simple application process that your SEO company should use, and then your company will receive verification by mail or phone. Your listing will appear in the Google search results within 4 to 6 weeks.

o   Yahoo Local (http://listings.local.yahoo.com/csubmit/index. php) – Yahoo local has a similar verification process but it takes a much shorter period of time. Within days of submission, your listing will appear in Yahoo Local's search directory.

o   Bing Local Listing Center (http://www.bing.com/local) – Formerly MSN local, listing with Bing gives your company coverage in the "big three" of the online search world.

o   Ask Business Center (http://city.ask.com) – Although Ask. com receives much fewer search queries per month than more popular search engines, submission to their City Business Listings can provide valuable backlinks.

INTERNET YELLOW PAGES

o   Yellow Book (http://www.yellowbook.com) – Yellow Book provides hard copy phone books and has delved into the online directory world as well. Listings on YellowBook.com frequently appear in the top 10 listings on Google.

o   SuperPages (http://www.superpages.com) – Getting listed with Super Pages is important because they supply listings to many portal sites, including MSN and About.com.

o   Yellow Pages (http://www.yellowpages.com) – Yellow Pages claims to have over 100 million local searches per month and offer a free online listing.

o   InfoUSA (http://www.infousa.com) – This is a business directory that power many high traffic directories. Getting listed in this directory can result in a lot of visibility.

LOCAL SEARCH ENGINES

o   Yelp (http://www.yelp.com) – Yelp was originally designed for the San Francisco area but it now houses listings for most metropolitan areas. The main focus is on listings and ratings. It's very Web 2.0 friendly and encourages user participation.

o   TrueLocal (http://www.truelocal.com) – TrueLocal has both free and paid listing options for businesses in the United States and Canada.

o   Local.com (http://www.local.com) – Local.com is a portal for a variety of different types of businesses and provides advertising for City Search, Verizon SuperPages.com and Yahoo! Insider merchant reviews. Getting a listing in this directory will open up your business to a lot of exposure.

This list of directories and local search engines just scratches the surface of the places that your local SEO company can submit your business directory listing. No matter how many places that they submit, be sure that they are submitting to a variety of locations including one from each major category.

SOCIAL MEDIA FOR LOCAL SEO

Finally, social media can provide your local SEO company with another important resource for optimizing your website and creating valuable backlinks. There are several social media sites that are equipped to link together local search results and popular social media applications. Your local SEO company can use these sites in order to boost your relevancy.

o   Flickr (http://www.flickr.com) offers meet up groups that your company can join as a member.

o   Outside.in (http://www.outside.in) can help your local SEO company put you in touch with local bloggers. Bloggers in your area are placed on a map depending on where they are blogging from. Your company can provide you with a list of local bloggers from this site or make contact on your behalf.

o   Placeblogger (http://www.placeblogger.com) offers similar services to Outside.in where you can browse for local bloggers.

RED FLAGS FOR OPTIMIZATION

✦   **Focusing on one form of optimization instead of another:** On page and off page optimization techniques will be used by professional local SEO companies to increase your chances of getting ranked for your local keywords. An overemphasis on one form of optimization over the other (using on page techniques rather than off page techniques or vice versa) will result in a shaky SEO strategy.

✦   **Ignoring advanced optimization techniques:** KML maps and hCard formats should be standard operating procedure for experience local SEO companies. There are many techniques for local SEO optimization that can readily be found online, and can be used by inexperienced SEO consultants to fake competency. The use of advanced techniques will help assure you that you are working with a company that has experience to back up their prices.

QUESTIONS TO ASK ABOUT OPTIMIZATION

✓ **What methods will you use to ensure that my website is going to be ranked well for my selected keyword terms?** Your local SEO company should respond with a comprehensive list of techniques that they will use that include both on page and off page optimization. They should describe editing your META content and on page content in order to optimize your site for your keywords. They should also describe their use of a variety of methods for increasing backlinks to your website and registering your website with appropriate directories.

✓ **What directories will your company use in order to list my company as a local business?** The list in this chapter covers some major directories that your local SEO company should utilize but they will probably also have some more specific suggestions. Be sure that the big three – Google, Yahoo and Bing – are in their plans for submission.

✓ **How will you incorporate social media into the optimization process?** An experienced local SEO company will know how to utilize social media in an effective way to boost the relevancy of your local keywords.

REPORTING

Your local SEO company should provide you with ongoing reports of how well your website is ranking for your selected keywords. The report should include keyword rankings, the number of backlinks and where those backlinks are coming from. The report will help you keep track of how well their efforts are going.

The first report should come shortly after they've taken you on as a client and will detail their plan for your site's optimization, similar to what was discussed in the SEO chapter of this book. The report will serve as a benchmark that you'll be able to use as the local SEO company begins working with your site. Following that the company should be in touch at least each month with updated statistics and feedback on how the optimization campaign is going.

RED FLAGS FOR REPORTING

+ **No initial report detailing the optimization plan:** You'll be able to tell quite a bit about the quality of service that an SEO company can give you once they send you this initial report. If a local SEO company is asking you to sign on the dotted line without presenting a report of their plans, look elsewhere for help.

+ **Reports that are inconsistent with your personal findings:** You can use Google Analytics or any other on page analytics tracker to see what backlinks and traffic your pages are receiving. If these statistics differ from the reports that your local SEO company is sending you, you need to question them about the discrepancy and possibly look elsewhere for your data.

QUESTIONS TO ASK ABOUT REPORTING

✓ **How will you update me on the progress of my optimization?** It should be clear from the start how your local SEO company will keep you up to date on the progress of your website's optimization. Asking this question to begin with will let you know what to expect.

✓ **Will your report include suggestions for future optimization plans?** The SEO company's monthly reports should be used as a tool to plan for future optimization steps, as well as a report of the progress that the company has achieved.

CONCLUSION

Local search engine optimization can get your company's website the local attention that it deserves. By combining general SEO knowledge to the unique needs of a local area, your selected local SEO company can get you more targeted traffic. Through onsite and offsite optimization methods they can increase your rankings and help you grow your business.

CHAPTER TEN:

AFFILIATE MARKETING
MANAGEMENT FIRMS

Running a successful affiliate marketing campaign requires a lot of work and maintenance. Affiliate marketing management can put a dedicated sales force in the palm of your hand, without the hassle of managing the program in house. By establishing and managing an affiliate program for you, your affiliate marketing company can increase your sales and grow your customer base.

An affiliate marketing management company will help you pay affiliates, also called publishers, when they refer sales and traffic to your website. Your affiliate marketing management company should be there to help every step of the way – from creating ad text to training affiliates to promote your product to paying affiliates for their sales. An expert affiliate management company makes it easy for you to put the power of the web publishing crowd to work for your business.

In the following chapter, you'll learn the key elements that should be part of any affiliate management company's process. Once you review these elements, you'll be able to evaluate if an affiliate marketing company has the capacity to help you achieve your goals with your affiliate program.

EFFECTIVE AFFILIATE MARKETING MANAGEMENT

An affiliate marketing company fulfills an important role in your online sales strategy. With the right strategy, network of publishers and payout system, your company can launch ahead of the competition and see results that you couldn't obtain on your own. An affiliate marketing management company should be well versed in the basic, intermediate and advanced principles of affiliate marketing.

On a broad level, an affiliate marketing company should help you identify your ideal affiliates, create advertising that will convert visitors, track the promotional activities of your affiliates and organize the payment of the affiliates. Affiliate marketing companies take all of the guess work out of establishing and running a successful affiliate marketing program. Your focus can remain on creating a terrific product or service.

GENERAL RED FLAGS FOR AN AFFILIATE MARKETING COMPANY

+ **"One size fits all" affiliate marketing management:** Your affiliate program should be geared toward your company's needs, budget and existing sales factors. A competent affiliate marketing management company should be able to personalize your affiliate marketing program to integrate into your company's existing approach to the niche and bring in new traffic from your target market.

+ **Lack of examples of successful affiliate marketing programs from well known companies:** Successful affiliate marketing companies should be able to point out specific, well known companies that they have helped in the past or are currently helping with affiliate marketing management. They should be able to show you testimonials from recognizable companies. They should not be using your company to practice their methods or learn the ropes. Seek out a company with an impressive client list so you can be sure you're in good hands. In addition, if they show you a list of specific clients, you can contact them independently and ask about the affiliate marketing management of the company in question.

+ **No standards for affiliate practices:** Once your advertising campaign is put together and handed over to affiliates, it should be promoted within a certain set of standards. If your affiliate marketing company does not require its affiliates to abide by these standards, your company may be put at risk. Affiliate marketing has come under a lot of fire in recent years for deceptive practices. The techniques that your affiliates use to promote your practices will reflect on your company. If the affiliate marketing company does not have affiliate practice standards, such as banning affiliates who use spam techniques, then look elsewhere. To check on the affiliate marketing standards for your potential company, look to their current client list and then try to locate affiliates who a promoting that company's program. If you notice any spamming, trademark infringement, false advertising or unethical methods in their promotion techniques, select a different affiliate marketing management company.

+ **Not providing affiliates with marketing materials:** If you want your affiliate marketing program to succeed, you need to give affiliates several tools to help them with promotion of your product or service. Graphics, banners, content, white papers and many other marketing tools can be used to make it easy for affiliates to get started promoting your site. The easier it is for affiliates, the more that affiliates will want to join your program and the better your program will do. Your affiliate marketing company should assist you with creating these important affiliate tools to increase your success.

+ **No automatic affiliate payouts:** Your affiliate marketing company should track affiliate sales and make appropriate payments immediately, at the end of the month or quarterly, depending on the terms of your affiliate program. It's important they compile all affiliate payouts into one invoice to you, rather than requiring you to manually approve affiliate payouts one by one.

GENERAL QUESTIONS TO ASK ABOUT AFFILIATE MARKETING MANAGEMENT

✓ **What type of needs analysis will you conduct before you begin working on our company's affiliate marketing program?** Not all affiliate marketing programs are created equally. Your affiliate marketing company should be able to establish parameters for your affiliate program based on your current traffic, your existing conversions and the nature of your market. The affiliate marketing program should only be created after determining what your target market wants and how affiliates can best reach them. For example, banner advertisements work with some niche markets while others respond better to written reviews. Only by analyzing your needs and the needs of your market can an affiliate marketing management company create a successful program that will attract affiliates and, therefore, converting traffic.

✓ **How will you select and train affiliates so they succeed with this program?** Support and grooming of affiliates is an essential part of having success with an affiliate marketing program. Your management company should have a specific procedure for courting affiliates and training them to making the most out of your affiliate program. In addition to the standard tools, such as banners, graphics and content, an affiliate management company may also use training videos, sales briefs, teleseminars and other advanced tools in order to attract and train top affiliates who can make a difference in your sales.

✓ **What methods do you use to track affiliate fraud?** When you engage affiliates you expect them to use above board methods for getting clicks and conversions. However, it's important that your affiliate marketing company use secure tracking programs to ensure that your affiliates are getting paid for actual traffic and sales and not those recorded by fraudulent means. A professional affiliate management company will have these systems in place to give you peace of mind.

NETWORK

The size of an affiliate marketing company's network is a good indication of how well they are managing the affiliate marketing campaigns for other companies. There is a difference between an affiliate marketing management company and the network that it uses to promote your program. You can sign up for an affiliate network on your own, but this does not guarantee that your affiliate program will be used by affiliates. It's only with the help of a capable affiliate marketing management team that you can see results from an affiliate network.

Affiliate networks come in two basic varieties – public and private. Private affiliate networks consist of the affiliate programs that are run by only one affiliate marketing management company. They engage groups of powerful affiliates by advertising their affiliate programs and utilize their client's websites to promote the affiliate program.

On the other hand, public affiliate networks are large collections of affiliate programs from many different companies. You may list your company's affiliate program on these affiliate networks yourself, but your affiliate program will not get much traction without the services of an affiliate marketing management company. Although there is more exposure with a large sized public affiliate network, it is also more difficult to become noticed in this environment.

A professional affiliate marketing management company should be able to advise you on which method is best for your company and your affiliate marketing program. They may be experienced with working within a larger network like Commission Junction (http://www.cj.com) and be able to get your affiliate program the notice it deserves.

They may have developed their own independent network of affiliate programs that is listed on their site, which will draw in sufficient affiliates in your niche to produce lots of referrals. No matter which method they advise, be sure that they have experience and results to back up the use of either one of these options.

A final method of networking is within the affiliate marketing community itself. There is a very active body of websites, blogs and forums that are dedicated to helping affiliate marketers find programs and effectively promote them. You should expect your affiliate marketing management company to have connections within these communities. Their connections will be essential in getting high quality affiliates to sign up for your program and effectively promote it. If members of your potential affiliate marketing management company speak at industry events, it's a good sign that they have the quality connections that you need.

RED FLAGS FOR NETWORK(S)

+ **Not using a network to promote your affiliate marketing program:** An affiliate marketing management company should have a network to promote your affiliate program, whether it is their own private network or they are experienced in working with one of the major networks.

+ **Offering no ongoing support once they have added your program to the affiliate network:** Having your affiliate program listed in an affiliate network is a bit like getting your business's phone number in the yellow pages. The program will be listed but it is unlikely it will garner any attention unless your affiliate marketing management company does some legwork in order to promote the program.

+ **Not paying affiliates on time:** One of the biggest red flags for an affiliate marketing network is delay or failure to pay out commissions owed to affiliates. Word spreads fast in the affiliate marketing community and if an affiliate marketing company is mistreating the affiliates, they will avoid it in the future no matter how promising your affiliate marketing program looks.

QUESTIONS TO ASK ABOUT THE NETWORK(S)

✓ **Where will my affiliate program be promoted?** A professional affiliate marketing management company will have several methods for promoting your affiliate program. In addition to using their own network of active affiliates, they may place your affiliate program on one of the popular affiliate marketing networks. On top of this, they should seek out active affiliates within your niche and create strategic partnerships that will be beneficial to both parties.

✓ **What steps will you take to ensure that my program gets participants?** Even if your affiliate marketing management company assists with getting your program listed, they should take further steps to make sure that it is promoted within the network and within affiliate marketing groups.

✓ **What kind of connections do you have within the affiliate marketing industry?** Look for mentions of quality affiliate marketing associations, such as Affiliate Summit, a long running affiliate marketing event and resource.

IMPLEMENTATION

The implementation of your affiliate marketing program will depend entirely upon the experience and knowledge of your affiliate marketing management company. Their method of implementation will have a profound effect on the success of your program. Besides their choice of network and their industry connections, there are many other factors that your affiliate marketing management company will have to decide on in order to create an effective affiliate marketing program.

Before your affiliate marketing management company begins implementation, they will help you define goals for your program and explain the options for achieving those results. They will develop a strategy based on your needs and aimed at maximizing your conversions. There are two specific phases to implementation – the development of the affiliate program and collecting affiliates.

The development of the affiliate program will include many different factors, all optimized for the best results for your particular program. Payment method is an important area of implementation. Depending on your goals for your affiliate program, you may want to use a "cost per sale" program or a "cost per action" program. These methods are performance based advertising models which reward affiliates for generating leads that take a specific action.

Cost per sale (CPS) is the more common of the two methods, with over 80% of current affiliate programs utilizing this process. With CPS, your company will pay affiliates a percentage of whatever sales they generate. Cost per action (CPA) pays affiliates when visitors fill out a form, click on a link or take some other type of specified action.

Another aspect of implementation to consider is multi-tier programs. Your affiliate marketing management company may utilize a multi-tier program for your affiliate program in order to encourage the first tier affiliates to find and recruit subsequent affiliates. In this model, your affiliate marketing management company will give a certain percentage per sale to the first level of affiliates. If one of those affiliates (Affiliate A) recruits another affiliate (Affiliate B), they will earn an additional percentage of income from Affiliate B's sale.

In addition to selecting the payment methods and tier structure, the implementation stage will also include identifying and creating the tools that will be necessary to establish your affiliate marketing program. These tools may include graphics,

Finally, you should seek an affiliate marketing management company that will implement your affiliate marketing campaign within a specified time frame. When you hire a company, they should be able to give you a firm deadline for the launch of the program. Professional affiliate marketing management companies are experienced enough to be able to estimate launch time frames so you can be confident that your program will be up and running quickly.

RED FLAGS FOR IMPLEMENTATION

+ **No clear plan for implementation:** Before you begin a project with your affiliate marketing management company, they should explain to you the steps they will be using to implement an affiliate marketing program. They should give you a series of steps they will take on your behalf, define your involvement and specify a time frame for completion.

+ **No advice on affiliate payment methods:** An experienced affiliate marketing management company will be able to advise your company on the best method for affiliate payout, whether it's CPS or CPA, and if a multi-tiered approach is best for your market. This is not a decision that you should have to make on your own.

QUESTIONS TO ASK ABOUT IMPLEMENTATION

✓ **What do you need from my company in order to get started?** In order to begin developing an affiliate marketing program, your affiliate marketing management company may need graphics, content and benefits lists of your product in order to proceed. They may need your input on these matters, or they may handle them entirely themselves. If you know you need to contract with your designer in order to create 125 X 125 affiliate ads ahead of time, the project implementation won't be held up.

✓ **When will my affiliate program be ready to run?** With affiliate program implementation, there should be a firm deadline for the end of the project. Ask early on and then hold your affiliate marketing management company to that date.

✓ **How soon will you begin contacting affiliates to promote the program?** Getting the affiliate marketing program ready to go is only half of the process. A qualified company will begin making connections with power affiliates to promote your program.

MONITORING

The monitoring process is essential to ensure that the affiliate marketing program is running well. Your management company should have a trustworthy system, which is normally a software tracking system, which will accurately record affiliate links, clicks and sales. The monitoring of the affiliate program should be reliable, not only for your company's sake but for the affiliates' sake as well. If your affiliates are not receiving commissions that are owed to them due to faulty monitoring processes, the word will spread quickly and your program will be blacklisted in the affiliate marketing community.

Another important aspect of monitoring is your management company's ability to monitor and police the use of your affiliate links by affiliates. This includes abiding by FCC rules for affiliate identification as well as avoiding spam and other fraudulent techniques. Your affiliate marketing management company can reduce the likelihood of these practices by screening affiliates or making connections with successful affiliates who already abide by these rules.

RED FLAGS FOR MONITORING

+ **Reports of late payment to affiliates:** By doing a simple search on Google, you'll be able to investigate whether or not an affiliate management company has a good relationship with the affiliate marketing community.

+ **No strategies for ensuring affiliates are promoting their links properly:** This is an essential part of affiliate marketing management that a company should be using. The success of your program depends on the affiliates and the methods they are using.

QUESTIONS TO ASK ABOUT MONITORING

✓ **What type of system to you use to track clicks and issue affiliate links?** Your management company should explain how their system works and how it ensures that affiliates are able to obtain their affiliate links and how it tracks commissions.

✓ **How reliable do you consider that monitoring system to be?** The affiliate marketing management company should have a very reliable system for monitoring click-throughs and commissions.

REPORTING

Once your affiliate marketing management company implements your program, you need to understand how well that program is doing on a consistent basis. Their reporting methods will be your only connection to the effectiveness of their program so it's important to understand how often you'll receive reports and what will be contained within those reports.

You should receive updates from your management company at least once per month, but probably a bit more frequently as the project comes together at the beginning. In the initial stages of creating an affiliate marketing program, your management company will be reporting to you in a less structured, more personal way. For example, they may send you a brief email each week explaining the progress on the project and if any input is needed from you. After this initial stage, you should expect to receive monthly reports on the program.

At the very minimum the reports should include click rates and other analytics information —like the time visitors spend on the site, how many clicked away, etc. Your management company will use industry standard reporting like this in addition to individual metrics of their choosing.

RED FLAGS FOR REPORTING

* **Lack of a sample report before project begins:** You should be able to obtain a sample report from your management company that details exactly what type of information you can expect on a consistent basis from your management company. You should ask about portions of the report that you may not understand and question the company about the reasons they are reporting on various metrics.

* **Incomplete analytics on the report:** When you evaluate the sample report, be sure that the report includes comprehensive tracking data. It should also include suggestions for future actions based on the reports.

QUESTIONS TO ASK ABOUT REPORTING

✓ **How often will I receive a report and what will it consist of?** Understanding this at the offset will help you know what to expect from the management company in the future.

✓ **How will the report be translated into action steps?** The affiliate marketing management company should have a clear plan for developing action steps based on the analytics data.

Analytics data by itself is not helpful. What matters are the steps that your affiliate marketing management company suggests based on the data.

OPTIMIZATION

Optimization is important in affiliate program management because it will help your affiliate program become better over time. Your management company should evaluate the affiliate program's progress and suggest changes for the campaign to improve performance.

The number one indicator of how well the affiliate program is going is the traffic and the conversions that your offer page is receiving. With the tracking and reporting methods that your management company is using, they should be able to pinpoint how well your affiliate program is doing and what changes need to be made.

As the results start coming in from your affiliate program, your management company should be able to find ways to adjust the campaign to improve performance. They will use the analytics information to determine what is going right with the program and what needs to be changed. It may be a case of the landing page needing to be optimized or the offer may need to be adjusted. Whatever the case is, your management company should be able to offer advice, make changes or connect you with services that can make the changes that you need.

RED FLAGS FOR OPTIMIZATION

+ **No support after the affiliate marketing campaign has been launched:** Your affiliate marketing management company should provide ongoing support for optimizing and improving the affiliate marketing campaign as the results come in. Based on their experience with other clients and the web analytics data, they should help you optimize the campaign so it is working best for you and your affiliates. Their job does not end when the affiliate program is launched.

+ **Service based on a time limit and not on the job completed:** Your affiliate marketing management company should follow through with the objectives that they set forth at the beginning of the process. Their job should be to provide support until that objective is met and not end their service based on a time limit of service. That being said, before you begin the service contract they should give you a time estimate for the completion of your service.

QUESTIONS TO ASK ABOUT OPTIMIZATION

✓ **How will you improve the performance of my affiliate marketing program over time?** Your management company should have specific answers on their optimization methods, as well as be able to show you through case studies and previous clients how their optimization methods improved the performance of your affiliate marketing program.

✓ **When will you consider the optimization of the program to be complete?** Although affiliate marketing campaigns can always be tweaked and improved upon, they should give you an estimate for meeting a specific goal. The goal and the time of completion should be agreed upon before getting started.

CONCLUSION

Affiliate marketing management can potentially make your company hundreds of thousands of dollars in sales over the coming months and years. By employing a dedicated team of affiliate marketers who promote your product and service over the Internet, you're positioning your company for massive growth in the future. Managing your own affiliate program can be time consuming, but by finding a team that will create, implement and manage your affiliate program by using the previous criteria can help your business grow.

CHAPTER ELEVEN:

MOBILE OPTIMIZATION COMPANIES

Mobile optimization is quickly becoming an important consideration for businesses of all types. More and more people are looking at webpages through their mobile devices. By the end of 2010 it's estimated that over 300 million people will be using mobile phones and PDAs to surf the Internet. Smart companies are embracing this usage and are seeking out ways to make pages viewable and usable by people who are using their mobile devices to access the web.

Mobile optimization companies will assist you with making sure that your sites are prepared for mobile users and mobile search engines. If your company has particular importance to mobile users, such as being a local retail business, then it will be especially important for you to embrace mobile optimization and gain the assistance of an experienced mobile optimization company. You want to have a site that is there for mobile users and easily read by them when they are "on the go" and looking for something in their area.

In this chapter, we'll go over how to approach optimization for the mobile market, including the similarities between standard SEO practices and mobile optimization practices. We'll go over what to look for in a mobile SEO provider and how to determine whether or not they're capable of doing a sufficient job for your company.

WHAT IS MOBILE OPTIMIZATION?

Mobile optimization is a combination of design principles and usability techniques that make your site easy to access for a mobile device user. Many of the principles of mobile optimization overlap with standard SEO optimization. However, there are other stylistic elements that will come into play with optimizing a site for viewing on a mobile screen. As a content publisher, you need to understand the differences between the way that a user browses through a website on their desk top and how they browse through a website on a mobile device.

For example, there are very real differences in the type of information that your mobile devices users will be looking for and your standard desktop Internet browsers will use. Mobile users will want to find information that meets their immediate requirements. They are looking for music, ringtones, directions or entertainment-related questions (such as movie show times). They also participate in social networking websites through their mobile devices.

Research shows that 50% of searches from mobile phones fall into seven categories. In addition to entertainment, directions and social networking, they look for sports, local knowledge, shopping and reference information. If your site deals with any of these types of information, mobile optimization is not only necessary but essential for your website's future success.

THE ESSENTIALS OF EFFECTIVE MOBILE OPTIMIZATION

Your mobile webpage design should follow a few basic principles for effectiveness. Your mobile optimization company should make sure that your website will look good on a mobile screen. Browse through the list of recommended mobile optimization vendors from topseos.com at http://www.topseos.com/rankings-of-best-mobile-optimization-companies.

Here are a few of the principles that your mobile optimization company should use in order to make sure your website is a good experience for mobile browsers. One of the biggest challenges in designing for the mobile screen is that there are several different sizes and layouts that a mobile device can use. The most standard sizes are:

o   128 x 160 pixels

o   176 x 220 pixels

o   240 x 320 pixels

o   320 x 480 pixels

Your mobile optimization company should develop a design that will function on all of these screen sizes so your website will be accessible to the biggest common denominator of people. Avoid mobile optimization companies that push adapting to the "latest and greatest" technology because not all mobile users are early adopters. Your goal is to be accessible to users that need your website and not exclude those who may have older technology.

Simplicity is the name of the game for mobile optimization design. The design team should avoid scripting languages, flash or other complex objects that will hurt the experience for the user. Users can't focus on detailed areas of design on such a small screen. Complex design elements can't be accessed easily from a small mobile screen. Asking your visitors to click on small links from a mobile device is like showing them the door – they aren't going to stick around. Mobile browsers and mobile users are much less lenient with design mistakes.

Your mobile optimization company should code your mobile site in XHTML. The strict standards that are required by XHTML design are ideal for mobile web page design. XHTML is also lean on code which will make it easy for mobile browsers to access your site. It will ensure correct display of your site on different screen resolutions. Keeping the design simple, making the interactions easy and putting the information your visitor needs right out there will make a huge difference in how effective your mobile site is.

If your website offers an appealing mobile experience, you'll become a relied upon resource for mobile users. A better user experience will result in more traffic and will lead to better search results in the mobile search engines. Following are the major areas of service that your mobile optimization company should address with their plan for your site.

GENERAL RED FLAGS FOR MOBILE OPTIMIZATION

+ **Approaching mobile optimization like a regular web design project:** You should seek out optimization firms that specialize in mobile design specifically in order to get a site that will work with mobile browsers. From the guidelines above, you can see that mobile screens have different necessities for usage.

+ **Creating an overly complicated user interface:** Touch screens, scrolling windows and the like can be hard to manage with complicated user interfaces, so your mobile optimization company should focus on a clean and simple design.

+ **Emphasizing the browsing capabilities of new mobile technology in their design:** Not all members of your target audience are early adopters of new technology. Mobile websites should be designed to meet the needs of a variety of types of visitors.

GENERAL QUESTIONS TO ASK ABOUT MOBILE OPTIMIZATION

✓ **How will your company optimize my website for mobile users?** They should present a comprehensive plan that shows their knowledge of the user's needs and your company's needs. They should demonstrate the basic design principles that are covered in this section as well as a dedication to providing an interactive user experience.

URL USE

The first area is URL optimization. Your mobile optimization company should be able to adjust the way that you allow mobile users to access your page. You can either have your page automatically generate under the same URLs as your regular webpage, or you can have dedicated mobile pages that are optimized for the mobile experience. You can also select to have subdomains that are accessible by mobile users that are part of your major site. Delivering the user to a stripped down version of the site through a redirect is important because it leads to a better user experience.

Your mobile optimization company should be able to make a suggestion of which method is going to be most advantageous to your particular needs. Using the same URL for both your mobile and desktop versions seems beneficial because your users do not have to remember two different URLs. If your mobile optimization company goes for this option, they will strip down or modify the content to meet the capabilities of your mobile browsers.

The concept of having only one site that is stripped down for mobile users may be tempting and is certainly less complicated from a coding point of view, but it doesn't create an optimal experience for your users. Users who are accessing your site through a mobile device have different needs and requirements than your desktop viewers.

Barring using the same domain, another option that may be offered is to develop an entirely new domain (i.e.: if your company's main website is www.yourcompanyname.com and your mobile website name is www.yourcompanyname-mobile.com). Having a separate domain can hurt your company's branding and confuse your users. You'll have to manage driving traffic toward two separate websites and list both on your business materials.

Your mobile optimization company may also choose to use a .mobi top level domain (which would read as www.yourcompanyname.mobi). The main disadvantage to a .mobi address is that it functions just like a new URL, along with a low page rank, no backlinks, etc. Mobi top level domains are becoming less necessary as websites are able to automatically detect what type of browsers your visitors are coming from and deliver a page in that format. However, their use is still much debated in mobile design circles. If your mobile optimization company insists upon using one, be sure that they can provide you with sufficient reason to do so.

The final and perhaps best option for mobile access is a specific subdomain for mobile users. Most mobile optimization companies will use a structure of http://mobile.yourwebsitename.com. This structure is easy to remember and it keeps your mobile site as part of your main domain. You can also opt for http://m.yourwebsitename.com. Having a mobile subdomain is the industry standard and is ideally what your mobile optimization company should use.

RED FLAGS FOR URL USE

- **Using a new URL or no difference in URL:** A brand new URL means that your optimization efforts and backlinking strategies may be split between two URLs. Your branding will also be harmed.

- **No difference in URL:** Using the same URL and having your mobile visitors visit an automatically stripped down version (based on the browsers standards and not your own) will create a poor user experience

QUESTIONS TO ASK ABOUT URL USAGE

- ✓ **Don't mobile browsers automatically configure sites to be seen on mobile phones?** This question is their opportunity to confirm to you that the automatic mobile versions of websites aren't going to be appropriate for your needs.

- ✓ **What type of URL do you recommend for mobile website versions?** Understanding your mobile optimization company's approach to URL usage is important before you move forward with their services. A professional mobile optimization company will be able to provide you with several good reasons why their choice for URL structure is the best.

COMPLIANT CODE

Different mobile devices use different operating platforms and therefore they have different browsers that your mobile optimization company will have to deal with. In addition, web pages for mobile browsing must be W3C compliant. W3C stands for the World Wide Web Consortium and it sets guidelines on how pages should be structured. They publish best practices and have a W3C verifier that will determine whether or not a site is compliant with the standards. Having familiarity and experience with the compliant code will assist your mobile optimization company with making sure your site displays the way it should to all types of mobile users.

The design team should optimize your interface for each browser and be sure that your site is meeting standard practice on all types of mobile browsers. This includes using a mobile validator to determine compliance with browsers but it can also include simple changes like using appropriate doc type and character encoding in page headers and keeping the site structure as simple as possible.

They can ensure this by testing your mobile site in one of the several standard mobile browsers. There are emulators available that can show your mobile optimization team how your design looks from a mobile phone. However, an emulator can only go so far. A professional mobile optimization company should go the extra step and have test phones on hand so they can easily see what your site will look like in a real world situation.

RED FLAGS FOR COMPLIANT CODE

- **No mention of W3C compliance:** Your mobile optimization company should be very familiar with the W3C standards and should emphasize their familiarity with them in their proposal.

- **Focusing on one type of mobile operating platform or downplaying older types of platforms:** All mobile users are important, and catering to only the top end phones is a good way to leave out part of your market. Any mobile optimization company that is trying to steer you away from complete coverage on a variety of phones is being lazy.

QUESTIONS TO ASK ABOUT COMPLIANT CODE

- ✓ **How will your company ensure that my site meets W3C compliance standards?** Your mobile optimization company should assure you that they are familiar with W3C standards and that they use the best practices outlined by the organization.

- ✓ **How does your company test for compliance across many platforms?** Avoid a company that merely relies upon emulators to test for compliance. This question will reveal whether or not the company has purchased test mobile devices in order to compliance.

NAVIGATION

Like other areas of mobile web page optimization, the navigation needs of the mobile user are very similar to the navigation needs of the desktop browser. Your mobile optimization company should make browsing easy for your visitors.

Buttons should be organized logically and should be intuitive. People are not going to adapt to your site, especially not on a mobile browser. They will expect that the information is organized in an understandable form. Because of the small screen size of mobile devices, it's important that they place the navigation below the content. This will give your visitors a more welcoming experience and leave more space in the first screen shot for helpful content. They won't have to scroll down to access important information.

Your mobile optimization company should use optimized jump links rather than traditional links right after the header. The optimized jump links should use your keywords to increase your mobile website's search engine optimization. Your mobile optimization company should consult with you on the importance of particular areas of your site so that those sections can be highlighted right at the top.

Other word links on the page should be kept very short. Your company should use abbreviations if possible in order to avoid taking up valuable screen space. Your entire mobile website should be organized so that important information is no more than three clicks away from your home page. Telephone numbers should be preceded by the "tel:" extension so that your users can make a call with one click. These guidelines will make navigation simple for users and will make your mobile website effective.

RED FLAGS FOR NAVIGATION

- **Setting up the site like a traditional web page:** Even a traditional webpage that is using best practices isn't appropriate for mobile users. Mobile users have different ways of accessing and searching for information and the navigation should be set up to meet those needs.

+ **Not using jump links:** Jump links make the site easier to navigate for the mobile user and they should be a standard part of mobile webpage design.

QUESTIONS TO ASK ABOUT NAVIGATION

✓ **What is your primary focus in designing websites for mobile users?** Their focus should be on the ease of use for the visitor combined with your company's needs. Usability is a huge factor in whether or not a site becomes a resource for the market.

✓ **Can you direct me to a mobile website that your company has designed?** Visiting a website that the mobile optimization company has designed through your mobile phone will allow you to experience how they handle navigation. Exploring what it's like to be a user on the site will tell you whether or not the company will provide an easy to navigate experience for your users.

INTEGRATION

Establishing the needs of your mobile website is important in working with a mobile optimization company. They should work closely with you to determine the purpose of your website and the interaction you'd like to offer to your mobile users. Your users should be presented with a uniform design and interface throughout your entire online experience – whether they are accessing your site via web or via their mobile device.

Design elements like color and basic layout should be consistent across the different websites. The goal of your mobile website should be similar to the goal of your regular website. For example, you should present your mobile users with content that isn't available on your regular website. The integration should be retrofitted to your main website and not the other way around.

RED FLAGS FOR INTEGRATION

+ **Pushing a re-design of your main website to fit the needs of your mobile website:** A professional mobile optimization company will work with your main website to integrate the design and functionality of your mobile website around what already exists. As long as your main website is doing well, there shouldn't be a need for an entire redesign. Even if your main website is very complex and includes elements that won't translate well to a mobile experience, your mobile optimization company should be able to take the basic elements and interpret those elements for a mobile audience.

+ **Using a design that is inconsistent with your main website:** Mobile website designers do not have free reign over the visual design of your mobile website. They need to comply with your existing colors and logo to create consistency with your main website.

QUESTIONS TO ASK ABOUT INTEGRATION

✓ **By looking at my current main website, what can you tell me about which elements will be used to provide integration between that experience and the mobile experience?** Your mobile optimization company should be able to go over some of the basic elements of their design that will help your main website and your mobile website be cohesive.

✓ **Will I be required to retrofit my existing website to be cohesive with the new design?** The only changes you will need to make to your existing website will be adding a link to the mobile site or adding a graphic that lets users know your website is accessible through a mobile device.

CURRENT

Mobile optimization is one of the most rapidly developing industries online and you should trust your mobile optimization efforts with a company who is staying on top of the changes in the marketplace. One of the most current developments is the previously mentioned W3C standards for mobile web development.

The goal of the W3C standards are to make browsing on the web with mobile devices reliable and accessible for the vast majority of mobile users. The basic best practices are similar to best practice for SEO, but they should be adapted for mobile users.

New best practices and forms of dynamic user interaction are emerging constantly in the mobile optimization market. For example, integrating social media tagging and social bookmarking tools is becoming more prevalent on mobile websites. An experienced mobile optimization company will be able to stay on top of these trends.

Search for a company that maintains an active blog, publishes press releases and articles on the industry or maintains white papers. These companies represent the thought leaders in the industry and are more likely to stay on top of the newest developments so they can stay current with their tactics.

RED FLAGS FOR BEING CURRENT

+ **A company that does not maintain an active presence in the market:** Your mobile optimization company should be publishing content to demonstrate their expertise. Whether this is done in the form of a blog, white papers or guest posts on industry blogs it doesn't matter, as long as the presence is there. Give preference to mobile optimization companies that have a presence in this area.

+ **A site plan that doesn't include the latest best practices:** You can check with the latest W3C standards and do some investigation on your own in order to familiarize yourself with current practices. Compare what you've learned against the plan that your mobile optimization company is suggesting to be sure that it is in line with current practices.

QUESTIONS TO ASK ABOUT BEING CURRENT

✓ **How does your company stay on top of the current developments in mobile optimization technology and practice?** Their answer should include ongoing education, conferences and participation in the general SEO and more specialized mobile optimization. They should give you a comprehensive list of their participation in the community as well as describe how they are testing and adapting their methods of optimization.

✓ **What industry resources do you use to stay on top of current developments?** By requiring them to name resources, you can do additional research on these resources to be sure that those resources are up to date.

CONCLUSION

Mobile optimization is field that is evolving rapidly. Look for a company that stays up on trends and is able to adapt to the newest technology, while still understanding the basics of usability for the end user. The guidelines in this chapter will help you separate the true mobile optimization experts from the general designers who are just claiming to have mobile experience. Looking at portfolios of previous work will be critical in determining whether or not a mobile optimization company has the experience that you need.

CHAPTER TWELVE:

VIRTUAL SPOKESPERSON SERVICES

Virtual spokesperson technology can help your website visitors learn more about your company and can increase your conversions significantly. Up until now, the one thing that has been missing from the Internet experience is the human touch. With a virtual spokesperson on your website, you can give your visitors a personal touch to their browsing experience and engage them in a cutting edge way.

Your visitors are exposed to an ever increasing amount of information online. They may land on your website after a series of clicks and need to be introduced to your site's purpose and what you have to offer. A virtual spokesperson can do all of this. In this chapter, you'll learn about this emerging technology and how a virtual spokesperson company can assist you with making your site a dynamic presence for your target market.

WHAT IS A VIRTUAL SPOKESPERSON?

A virtual spokesperson is an online person or character that is authorized to speak with customers on behalf of your customer. The most common use of a virtual spokesperson is to greet your visitor as they arrive at your site. Here's an example from ISpeakVideo.com.

(source: http://www.ispeakvideo.com)

They engage in a one way dialogue with your website visitors and can encourage the visitors to take conversion steps. Virtual spokespeople on your website can:

o   Capture the attention of the audience immediately.

o   Showcase products or services on your website.

o   Explain your unique selling proposition.

o   Ask visitors to fill out a form.

o   Retain visitors for longer periods of time.

The possibilities are endless and a professional virtual spokesperson company can help you engage your visitors in new ways. Since the use of this technology is still relatively new you can sky rocket your website past your competitors while they are still trying to piece together how this works. Your virtual spokesperson can give your website a big "wow" factor.

The virtual spokesperson video will pop up into the landing page frame as the visitor hits your page, or after they've been browsing through your site for a certain period of time. They can deliver more information to your website visitors in 30 to 60 seconds than if your visitor were browsing through two or three of the pages. People can understand things better when they listen rather than read

USING A VIRTUAL SPOKESPERSON EFFECTIVELY

There are many aspects you should look for in the services of a virtual spokesperson company. The following categories represent areas that you should look at when evaluating a potential virtual spokesperson company. In addition to the categories listed here, you can look at the topseos.com recommendations for virtual spokesperson companies at http://www.topseos.com/rankings-of-best-virtual-spokesperson-companies.

VISUAL APPEAL

There's no point in having a virtual spokesperson if their presence is disturbing to your visitors. The point of having a virtual spokesperson is to have your visitors want to stay and if the virtual spokesperson is unappealing, your bounce rate will increase and you'll lose out on converting your visitors.

Look for a virtual spokesperson interface that will be appealing to your audience. Your potential video presentation company should be able to show you several different examples of videos that they have produced. Analyze the approach of the virtual spokesperson on the samples pages. Does the presenter create an emotional and personal connection with the audience? Is the communication clear and easy to follow? How is the quality of the video? Although your virtual spokesperson company will create a customized presentation for your company, it's important to look at samples first before you commit to working with a company.

VIDEO FEATURES

There are many features that can be integrated into a virtual spokesperson presentation that will increase its effectiveness. Features can include something as simple as the placement of the video spokesperson to more complex features like floating video presentations. A quality virtual spokesperson company will be able to offer you videos for several different needs. Some examples include:

o   A corporate presentation with graphics, text and animation.

o   Presenting a different video each time the visitor reaches the page.

o   Presenting a different video depending on where your visitor clicks on the page.

o   Presenting your FAQ section in video form.

The company should have the capability to provide you with the quality and style of video that you need. They should offer several in video features like the type of display, scrolling video, background music and interactive elements.

The display options should be vast for your video presentation. You should have a choice of sizes, a choice of the viewable area of the actor (full body, close up, ¾ body shot) and placement of the video on the page.

You need to look at additional features of the virtual spokesperson company such as customer support, script development, reporting and tracking. A professional virtual spokesperson company should be with you every step of the way to ensure that the video meets your company's needs and will approach and convert your visitors.

SIZE OF ACTOR NETWORK

Ideally a virtual spokesperson company will allow you a variety of models to choose from for your video needs. The more models, the better, especially if a virtual spokesperson company has been used frequently in your industry. If the company is being used frequently and there are only a few choices of actors, your website's presentation will look less unique.

Many virtual spokesperson companies offer a wide variety of actors for you to choose from. This will make it easier for you to find the right actor to present your business. You might need a fun and casual presenter or someone who is older and distinguished. A professional company will be able to meet your needs with a stable of actors to choose from.

You should also be able to have your own staff member or CEO act as the spokesperson for your company. If you want to go this route look for a virtual spokesperson company that has studios in major cities so your staff member can have the video professionally shot and edited by the company.

Alternatively, a company may provide onsite recording where they come to you in order to

INTEGRATION

Your video presentation won't be effective if it isn't integrated properly into your website. The integration step is crucial to using a virtual spokesperson effectively. The virtual spokesperson company should be able to work with your webmaster to input the code into your website seamlessly.

It's important that the video appears on the website in a location that is viewable no matter what size screen your viewer is using. The best place for the virtual spokesperson is to hover over the text that relates to what they are talking about or in a blank spot that has no distracting graphics or important text. The video should be placed in a cleared out area of the site that doesn't include any important links, opt in boxes or calls to action.

You can place your video into your website at several different places. The home page is popular for a welcome message. You can also place the virtual spokesperson on lead generation forms, e-commerce check out, call to action and thank you pages, in banner advertising and email campaigns and as part of demonstration and training tutorials.

Your virtual spokesperson company should work with you to define your goals for their services and help you determine where the best place would be for your spokesperson. They should be able to integrate the video into your website at any point you need it to be.

REVIEW PROCESS

Before your video is placed live your website, you should have final approval on the content and quality of the website. Your virtual spokesperson company should not place anything on your website before you have approval. Be sure to ask questions about their approval policy and see if the company stipulates that you can have a re-shoot on the video if you aren't satisfied with the results.

RED FLAGS FOR VIRTUAL SPOKESPERSON SERVICES

+ **Small amount of sample videos:** If your virtual spokesperson company does not have many sample videos to show you,

you can't be sure of their expertise. Look for a wide variety of samples that showcase the virtual spokesperson company's familiarity with the format.

+ **Poor video quality:** From the sample videos, you should be able to tell if the quality is up to par with what you want to present on your site. As you are evaluating sample videos, remember that your virtual spokesperson will be a representation of your business and poor quality video will offset the benefits of having the video presentation.

+ **Narrow list of video features:** Welcome message videos on the home page are the most common use of virtual spokesperson technology, but this does not mean that it is necessarily the most appropriate usage for your website. Seek out a company that offers a variety of features so that your virtual spokesperson can be customized exactly to your needs.

+ **Limited actor pool:** We've already gone over the benefits of having a wide variety of actors to choose from and a limited actor pool is a bad sign for a company's ability to give you what you need.

+ **No script input:** Most virtual spokesperson companies will have you create your own script for them to work with, but it's important that they are able to provide professional feedback and advice. Since they have experience in this arena they should be working with you to ensure that your script will meet the needs of your audience in order to convert more visitors.

+ **No re-shoots:** If you aren't happy with the quality of your video, you should be able to get your video redone at no additional charge.

QUESTIONS TO ASK ABOUT VIRTUAL SPOKESPERSON SERVICES

✓ **Can I see samples of your virtual spokesperson services in a variety of situations?** Any reputable company will have several samples for you to review.

✓ **What type of customization can you offer for my virtual spokesperson video?** Look for a company that can help you create a customized video that is unique to your website and your needs.

✓ **Can you offer multiple videos on one page?** The ability to create several videos on the same page depending on how many visits the visitor has made or the places they click on the page will greatly increase the effectiveness of your video.

CONCLUSION

Virtual spokesperson software is not necessary for all types of websites, but if you feel your website would benefit from a virtual spokesperson, these criteria will help you choose the right company. Looking at the sample videos and getting details on the company's features will assist you with finding a company that can meet your artistic and technical needs.

CHAPTER THIRTEEN:

HOSTING COMPANIES

Your online business can't function without a website, and a key part of having your business website is your hosting service. Your hosting company will provide space for your website's data files and handle your payment and delivery system, if you sell products from your website. In the following chapter, we'll take a look at some of the major issues concerning hosting and how to determine if you're using a hosting company with the best service, features and reliability. Selecting a hosting company is no small matter and the selection criteria for a commercial website will be important in finding the right option for your business.

Hosting companies come in many different varieties, and there is a comprehensive list at topseos.com located at http://www.topseos.com/rankings-of-best-hosting-companies.

THE CHARACTERISTICS OF RELIABLE WEB HOSTING

There is a lot of competition out there for your hosting business so it's important to make your selection wisely. This isn't an area that you should shop for only by price. The effectiveness of your website will be a direct reflection of the quality of the hosting you purchase. The following areas are important in evaluating a web host.

VARIANCE

Variance is an important quality to have in a hosting company because as your business grows you'll need various hosting options. The first option that you need to look at is monthly bandwidth. A reliable web hosting company should give you sufficient monthly bandwidth so that data is easily transferred to your website visitors. The bandwidth is the amount of data that goes in and out of your server space when website visitors reach your content, watch a video or otherwise interact with your site. Your hosting company will likely have several levels of bandwidth space for different hosting plans.

Do not believe commercial webhosts that promise unlimited bandwidth. The hosting company has to pay for bandwidth and if your website uses a lot of bandwidth, they will likely charge you for it. Exceeding bandwidth will result in large hosting bills. Be sure that your commercial webhost specifies the amount of bandwidth your website is allowed.

As your site grows in popularity, your traffic will increase. Be sure that your hosting company has options for increasing your bandwidth without exorbitant fees. You should also check that your hosting company offers multiple domain hosting, in case your company wants to expand its web presence in the future to include multiple properties. Becoming familiar with your options before you settle on hosting for your site will help your website grow with time.

Disk space is another area of variance between hosting plans. The disk space is the total amount of data space that your website can occupy on the hosting company's servers. Generally, hosting providers give your company plenty of hosting space for your needs but as your site grows you should be able to expand your disk space options.

Look for a hosting company that provides both shared and dedicated servers to meet your needs. A shared server means that you are sharing the server unit with several other companies, which means you only have a limited amount of space. With a dedicated server, your company's website is the only one on the physical server so you have unlimited growth potential.

UPTIME

Having a reliable web host that is up and running over 99.5% of the time is ideal. There is no excuse to work with a hosting company that does not provided comparable figures for up time. Look for a host that will offer a refund or discount for service if they fall below that percentage. Most reputable hosting programs will offer a guarantee of service for uptime, which may be hard to enforce from your end, but the presence of a guarantee will give the hosting company incentive to make sure their servers are running all of the time.

FEATURES

The first feature to look for is the type of operating system and server type. This will make the difference in how your website is able to be used. For example, if your website needs to use ASP programs you will need a Windows server. However, a Unix system running on an Apache server will work for most other types of websites, including PHP based websites. Apache servers also let your company easily add error pages, image protection, blocking email harvesters and block IP addresses. Apache servers are the industry norm, and unless your website needs specific ASP programs, you should seek out a host that uses an Apache platform.

The features of your hosting package are essential to your ability to create an effective website. Some of the main features you should look for include FTP, PHP PERL, SSI, .HTACCESS, Telnet, SSH, MySQL and Crontabs. These features offer different options for accessing your site and give your web team complete control over your hosting.

Access to these features is preferable to hosting packages that do not allow the installation of PHP or Perl scripts without approval. This will slow down your web development and may cause problems with the use of your site. .HTACCESS is important to help your web development team create customized error pages and take steps to protect your site from bandwidth theft and hot linking.

Telnet and SHH is essential for testing scripts and maintaining databases. MySQL is required for content management systems and blogs. Crontabs can be used to run programs at certain times of the day. Your hosting package should offer use of all of these types of programs in order to help your web development company create and manage your site.

Look for a web host that offers a log file analysis tool. Webalizer is the industry standard for analyzing log files, but there are also other options. No matter what the choice your hosting company uses, it's important to have the ability to export log files onto your computer so you can utilize your own statistics reporting tools.

Another important feature is having automatic backups of your data so if something goes wrong with your hosting, you can recover the data and begin again. Your hosting company should tell you what type of data is backed up (files, databases and/or server settings) and how often these backups are made. It's also important to find out where the backups will be stored (it should be off server or offsite).

A number of separate FTP access points is helpful if there will be many collaborators working on your website at once. With multiple FTP accounts your web development team will be able to collaborate efficiently and even pull in other third party developers without revealing your company's personal credentials.

Another helpful feature for your hosting company is integration with publishing platforms, like WordPress or Movable Type. The publishing platforms can be used to quickly establish a content management system or add a blog to your website if one is needed in the future.

Having access to a control panel is essential to developing your website. Every web host will call the control panel something slightly different, but essentially it allows you to manage the data that is stored on the hosting. If a company requires you to speak with technical support to do things as basic as changing passwords or adding/deleting email accounts, they will waste your time. Control panel access is an absolute must for business hosting.

Not only should your hosting company provide a control panel, but it should be clear and easy to understand. Looking at a screenshot of the control panel will let you judge for yourself whether or not the control panel will provide the features, and ease of use, that you need.

If your website requires a shopping cart, you should seek a hosting company that provides secure server (SSL) shopping cart integration. The hosting company will normally charge a slightly higher price for this option, but it is well worth it if your hosting can seamlessly integrate a secure shopping cart into your website. An SSL shopping cart is required if you plan to collect credit card payments on your site.

Email addresses, POP3 access and mail forwarding will allow you to have an email address (or several addresses) attached to your domain. Be sure that you can easily access the email by forwarding it to an email software client or your existing email address. Autoresponders are another key feature to look for in this department.

STABILITY

Stability is an important factor in a reliable web host. You want to know that the hosting company will be around for years to come. Switching your data between web hosts is cumbersome and produces problems with visitors to your site. You can avoid this problem by finding a web host that will be around for years to come.

Although it's impossible to predict the future, you can tell a few things about the stability of a hosting company by doing some investigative work. Look for the hosting company in the better business bureau and do a general web search for the company name. Industry reports, consumer reports and top 10 lists can be a great source of determining the company's history and reliability. You should also look at what others are saying online about a particular webhost. The reputation of the webhost can easily be determined by looking at reviews and reading comments on forums. Don't let one negative review sway you, or one positive review for that matter, look at the overall impression from a variety of sources. Generally speaking, the longer that a web host has been around, the more likely it will be around in the future.

Make sure you are working with a web hosting company directly and not a reseller. Some hosting companies are actually leasing web server space from another web hosting company. They are reselling the extra space at a profit. A reseller may not be as well versed in the system they are selling as the actual web host will be. It will take longer to get assistance with a reselling provider account, because they will have to transfer their support requests to the main company and then get back to you.

That being said, there are a number of reliable web hosts that are actually resellers. It all depends on their track record and level of service. Some resellers actually sell packages at a lower rate than the original hosting company so it can be worth it to investigate a reselling company. Just be sure to thoroughly check out both the reseller and the hosting company that is providing the service.

SERVICE

The service of a web host will help your company's website run smoothly. You're employing a hosting company to make your online business life easier. With the proper level of service, you can be sure that your webhosting company will be there to support your website's development.

One of the main areas of service is in their payment plan options. A web host with normally offer a monthly, quarterly or annual payment plan as options for your business website hosting. Avoid a company that forces you to prepay for an annual plan. Paying monthly on a new website will allow you to test out the hosting company's service and see if you want to keep them. If you are forced to prepare for the entire year, you won't have this level of freedom. Once you test out the reliability of the web host, you can save money by paying quarterly or annually.

Technical support is essential when you're working with a hosting company. Your hosting company should be there for technical support 24/7 with no exceptions. Problems with your website may arise at inconvenient times and it's essential that your hosting be there to support you no matter what your needs are. Live support that can be obtained quickly is essential to your site's success. You also need to verify that the support providers are not just salesmen but actually technically knowledgeable about the hosting they are providing. |

RED FLAGS FOR HOSTING

+ **Meaningless "unlimited" promises (ex: "unlimited bandwidth" and "unlimited storage"):** This is a red flag because giving unlimited bandwidth and unlimited storage are sales techniques and not based in reality. If your web hosting company promises these unlimited services, you may be in for a rude awakening if you actually approach the limits. It's much more prudent to be familiar with the concrete bandwidth and storage limits that you can deal with.

+ **Free hosting:** As a business, you need to provide a professional experience for your potential customers. Free hosting may help save the bottom line, but a free web host will put banners advertising their company on your website. This detracts from your sales message and makes your site look amateur.

+ **Less than 99.5% uptime:** When it comes to web hosting, even 99% up time is too little to work with. Just .05% makes a difference in wasted resources and wasted time. Search until you find a reliable web host with the highest uptime possible.

◆ **Short length of time in business:** Web hosting companies are easy to set up and hard to maintain. You shouldn't trust your web presence with a company that has only been in business for a few months. Do your due diligence on your potential hosting companies and find one that has been in business for several years and looks like it will have stability well into the future.

◆ **No access to the control panel:** Your web development team needs the freedom to access the control panel and make necessary changes to your website's functionality. Don't tie their hands by selecting a company that does not provide this all important access.

◆ **No 24/7 support:** The Internet doesn't sleep, and you should be able to get access to technical support whenever you need it. Test out your web hosting company's claims of 24/7 support by sending an email on the weekend and seeing how long it takes for them to respond. Consider switching hosts if you don't get a timely response. 24/7 access to live chat or telephone help is preferable.

CONCLUSION

Hosting is the foundation of your online presence, so research your options thoroughly before you invest in a hosting package. It will be difficult to move your site once it's been established on a particular hosting plan. Use the guidelines above as well as the rankings at topseos.com to find a hosting company that you can rely on now and well into the future.

CHAPTER FOURTEEN:

PSD TO HTML CONVERSION SERVICES

There's no discounting the importance of having a good looking website to greet your visitors when they do a search for your website. Often times your design team may come up with a visually appealing design that needs to be changed into a fully functioning website. This is where PSD to HTML conversion comes in.

A professional PSD to HMTL conversion company will be able to take the graphic designer's concept and enhance it so that it can function as a website. Having a beautiful design is worthless if the coding isn't done correctly. If you want a rich and visually attractive design, a PSD to HTML conversion company can help you get there. Following are the qualities that you should look for in a professional PSD to HTML company. For more specific recommendations, see the rankings at http://www.topseos.com/rankings-of-best-psd-to-html-companies.

PROFESSIONAL PSD TO HTML COMPANY PRACTICES

Like many other Internet related service industries, the PSD to HTML marketplace is filled with a wide range of skilled providers. There are providers on the low end that offer rock bottom prices (along with less than professional skills) and there are providers who charge higher rates because they deliver higher quality. Besides the difference in price, there are other factors that you can use to compare PSD to HTML companies to determine which the best choice is for you.

Shopping on price alone can lead to disappointing results. While there is something to be said for shopping for the best deal, the following criteria will help you determine what the best deal really is. The categories below should be used in your evaluation of PSD to HMTL companies so you can tell whether or not their services are worth your time and money.

CLEAN CONVERSION

The success of a quality PSD to HTML conversion is making the workable website look the same as the PSD file. A professional company should be able to make a seamless transition between the original file and the finished product. The CSS classes and images should be intuitively named so that the code is easy to access in the future.

Clean conversion is important because a cleanly coded website is easier to access for your customers and any people who will be working with your website in the future. Although you may go back to the same PSD to HTML company in the future for further development, a professional company will make your design clean enough so that any skilled designer can access your files at a later date.

Professional PSD to HTML conversion companies will create a professional look and feel for your website or your blog. Their coding will help search engines understand your website which can lead to increase rankings. Using xHTML/CSS coding is very much "in style" now in the world of web design and a professional company will make your site look on top of the market.

A professional company will build the code from scratch to make XHTML and CSS files that translate your visual design into a usable website interface. They should use hand coding rather than working with a WYSIWYG (what you see is what you get) software. WYSIWYG software produces unwanted code that can slow down the load time and effectiveness of your website. Ideally, the company should separate the structure, presentation and the behavior of your website into separate codes (XHTML, CSS and JavaScript respectively).

The end goal of your PSD to HTML company should be to create a website with fresh look with a fully functioning interface. Their conversion services should also include extensive testing on their design once it's been coded. This will ensure that the coding works and your site is ready for use. Evaluate their portfolio of previous projects to see if the end result is pleasing to the eye and fully functional. This will be the best way to measure their ability to produce a clean conversion.

STANDARDS COMPLIANCE

Many people who do business online aren't aware that there are standards of compliance for websites. The most common set of standards is issued by the World Wide Web Consortium (W3C). Creating a website that is W3C compliant is essential in order to make sure that your website is read correctly by the search engines and are understood by your visitors.

The W3C puts forth "best practices" for web design that your PSD to HTML company should be following. Although there are specific details for each type of website, there are general guidelines that your PSD to HTML company will follow including:

o   Converting to clean HTML.

o   Optimizing the images for keywords and visual look.

o   Implementing JavaScript seamlessly into the design.

o   Developing CSS code without tables.

o   Compatibility with several browsers.

Compliance will help your website operate more fully online and can also help boost your search engine operation. Make sure that your PSD to HTML company operates within W3C compliance standards and meets the needs of the search engines and your users.

TURNAROUND TIME

Turn around time is an essential factor to look at when selecting a PSD to HTML conversion team to create your website. The standard turnaround time can be anywhere from a few days to two weeks depending on the complexity of your design and the features needed. Look for a company that guarantees their results within a certain period of time and gives a firm deadline for delivery of the work.

Getting compliance with the deadline is more important than a short turnaround time. When it comes to comparing companies, you should go with a company that has a reasonable turnaround time and quality work rather than one that promises 24 to 48 hour delivery. It's better to have a designer that will take more time and product quality work rather than deliver a "fast food" style website. If your PSD design is overly complex, this is even more important.

INTEGRATION

A professional PSD to HTML design team will be able to translate your visual design into a variety of different content management systems depending on your company's needs. For example, you may want to have your design translated into a purely HTML page or you may want to utilize the WordPress platform as a content management system.

Most PSD to HTML companies will be able to integrate a number of third party software programs that will help your website work for your market. These software programs can include:

o   CMS Made Simple

o   CubeCart

o   Drupal

o   Joomla

o   MagnetoCommerce

o   Miva

o   MODx

o   NetSuite

o   osCommerce

o   Pligg

o   Shopify

o   X-Cart

Look at the portfolio of the PSD to HTML designers and ask questions about their integration capabilities. If you're not sure which type of CMS you'll be using, ask for advice from the design team. By communicating your functional needs they should be able to come up with a CMS that will serve your purposes and then be able to connect your visual design with that CMS.

PRIVACY

It's essential that your PSD to HTML design company protect your privacy while they are working with your site. The information that they'll have access to will be critical to your businesses success and you should take measures to protect your information. Your PSD to HTML company should be completely okay with signing a Non Discloser agreement (NDA) in order to ensure that they can't use some information that they'll have access to.

Don't allow the PSD to HTML company to access your company's information unless they've signed the proper paperwork. This shouldn't be a problem for a professional company.

RED FLAGS FOR PSD TO HTML

+   **Lack of samples of previous work:** Don't take a chance on an inexperienced conversion team. The company you hire should have an extensive body of work on display on their website or at your request.

+   **Conversion using a WYSIWYG editor:** Be sure to ask your potential conversion company about their methods of conversion. They should emphasize hand coding in their conversion practice and not rely on a WYSIWYG editor.

+ **No delivery date:** Turnaround time for the PSD to HTML project should be settled before the project begins and before you are expected to share your private login information.

+ **No mention of standards compliance:** W3C compliance is the industry standard. A PSD to HTML company should clearly state on their website that they adhere to these standards and can create a website that is compatible with a variety of web browsers.

+ **Pressure to share login information before they have been officially hired:** Do not use a company that pressures you to share your login information to "check things out" before they officially start on your project.

+ **No integration with necessary third party programs:** Be clear about what you need your site to integrate with and don't let the PSD to HTML company pressure you to use a different solution.

CONCLUSION

PSD to HTML conversion can give your site the extra visual touch and smooth usability that most website users are looking for. The functionality of your websites relies upon this process being done correctly, so investigate your options thoroughly. Look at sample designs, speak with previous customers and make sure everything is spelled out in a contract before getting started.

CHAPTER FIFTEEN:

PRESS RELEASE DISTRIBUTION CHANNELS

Press releases can be powerful weapons in your company's attempts to gain search engine rankings, get more visitors to your site and increase prestige within your market. With a skilled press release distribution company, your company can get widespread exposure. Press releases can also help optimize your website for specific keyword phrases. Press release submission companies take your press release and put them in front of journalists, bloggers and people who can make a difference in your business.

In this chapter you'll gain insight into how traditional press releases have been transformed by the Internet. Your press release service can boost your rankings and get you exposure that you couldn't on your own. Although there are many free press release distribution services out there, putting your message in the hands of a capable professional service will help your press release have more impact.

In addition to the qualities discussed below, you need to search for a quality content provider or press release writing training because some areas, like optimization, are primarily on the shoulders of the company creating the press release. See the Content Creation chapter of this book for criteria to evaluate content companies. Further analysis of press release distribution providers can be done at topseos.com, which has been evaluating SEO companies since 2002. Press Release Distributors are featured at http://www.topseos.com/rankings-of-best-press-release-distribution-companies.

WHAT IS AN SEO PRESS RELEASE?

Press releases have long been a staple in newspaper and magazine offices however; they have developed new purpose online as tools for both publicity and search engine optimization purposes. Old school press releases were designed to attract the attention of news editors who would pass the news on in their publications. Today, press releases are accessible by everyone through search engines

SEO press releases increase the visibility of your press releases for the media and your customers alike. With the right submission strategy, a press release service can get your news picked up by Google and Yahoo search engine news networks, which will greatly increase your exposure. In some cases, your news item will appear on the first page of the news results for those search engines.

Having a press release created and distributed for your company can bring five major benefits.

o   Organic traffic – Google, Yahoo and Bing all rank and list press releases within their organic search results, as well as their targeted news results. By using keyword optimization methods, your press release will be part of search engine results. It will also be listed on news aggregator websites which can give you another outlet for traffic.

o   Link building – Links are of utmost importance to the long term viability of your website. By using a press release optimized for your keywords and linking back to your website, you increase your chances of being indexed for those keywords.

o   Reputation Management – Multiple press releases distributed through a press release service can replace any bad press that your company has received on the first page of search engine results.

o   Exposure to the Web 2.0 Audience – Many bloggers subscribe to online press releases services and receive press releases each day related to their interests. If your press release will make a good story for their blog, they will create a post about it and link to either your press release or your website. This can create additional backlinks and industry exposure.

o   Traditional media exposure – Online press releases are used by traditional journalists to source story ideas. If your press release appeals to them, they will contact you to create an original story about your business. The exposure from traditional media can increase your traffic and sales.

(Mody)

Case Study: PatioConnections.com

Press Release Distribution Service: Online PR News (http://www.onlineprnews.com)

PatioConnections.com had been struggling with driving traffic to their site through various sources such as PPC and SEO with what owner Dean Geyer called 'very poor results.' As a new site (less than 6 months old) in a competitive market with more than 2 million results for many of his search terms, the website was not ranking organically for its target keyword terms. While they continued to work on a long term SEO strategy, the company needed to achieve first page visibility at an affordable rate -- quickly.

THE BASIC COMPONENTS OF PRESS RELEASE DISTRIBUTION

A quality press release service can get your company instant exposure online. There are several key elements that set a press release service apart from the rest. These elements have a direct effect on how successful your press release submission process will be. Before we delve into these elements you should know that most press release services require that you have press releases written before using their services. Some may recommend a press release writing service, but most will expect you to have your press release ready to go. For more information on hiring a press release writer, please see the Content Creation chapter of this book.

SUBMISSION PROCESS

Before you pay for a service you need to investigate their submission process so you are familiar with their requirements. The submission process should be clear and easy to follow. The point of employing the services of a press release distribution network is to make things easy on your company. Complicated instructions, endless loopholes or narrow submission guidelines will not serve your needs.

A professional press release distribution company will make the process as easy as possible for you. Ideally, you should be able to enter your headline, the body of your press release, your keywords and your image easily. The press release distribution service should have intuitive controls and allow you to select what industry your press release is appropriate for. In addition, you should be able to select the region or theme for your press release.

Many professional press release distribution networks can make it simple for you to add anchor text and links. Some of the best offer formatting help with hyperlinks and text formatting so you can be sure your press releases are optimized for SEO. Look for a service that offers help with making the most of your press release.

OPTIMIZATION

Unlike the other areas of press release distribution, optimization is primarily the responsibility of the business owner, a member of the staff or a hired press release writer. Unless a press release service offers writing services, optimization of the press release will be firmly on the shoulders. Optimization is the key in making sure that your press release gets in front of its intended audience.

CASE STUDY: *PatioConnections.com*

Submission Process and Optimization:

As a business owner new to online marketing, Dean Geyer of PatioConnections. com was feeling frustrated with the steep learning curve of internet marketing and gaining online visibility. After learning that online press releases provided a good source for both links and exposure, he decided to try his hand at submitting a press release at Online PR News' $49 level to test the results. PatioConnections.com's press release was received by one of Online PR News's editors trained in search engine optimization and PR. While the press release was good, the editor realized there were several things that could be tweaked to enhance both the value of the press release from an SEO perspective and a viral perspective. The editor offered specific instruction on how to make these changes, such as choosing more effective keywords and optimizing the title. Dean quickly made the changes and the press release was published.

If you want to issue a press release, but aren't sure that you have anything newsworthy to say there are several ideas that you can use.

o   New product or service: This type of press release is only effective if it is framed as a solution to an existing problem that people are concerned about. For example, the focus of the press release would be the increase in carpal tunnel related industries. The press release would also announce the introduction of your

o   Results of a customer survey: Survey results give you the chance to brand your company but also release information that is relevant and important to the industry as a whole. This is more likely to be picked up and distributed than a new product or service announcement.

o   Announcement of an article series or report available for publishing: Article series or reports will attract your target market, bloggers and others in your industry. A press release will let those groups know that the documents are available and that they can be used for distribution. It will increase traffic to your site and grow your opt in list.

o   Announcement of sponsoring a workshop or webinar: Similar to offering an article series or report, this type of press release has something to offer your market above and beyond a simple promotional piece.

o   Announcing your availability (or a staff member's availability) to speak on particular subjects: This type of press release will brand your company and open up additional opportunities for you or your staff member to speak on industry specific topics.

o   Issuing a statement of position on an important issue: Occasionally, there may be important issues that come up that directly reflect your industry. Journalists love these types of press releases because they provide a great source of research for articles on these issues. They will often contact people based on their "position" press releases.

o   Statements on future business trends: Similar to a statement of position, this type of press release puts your company at the forefront of industry contacts.

(Lautenslager)

Once a press release distribution service receives your news worthy item, they will determine whether or not it is optimized for release. If they offer editing services, they will walk you or your staff through the process of making sure that the press release is newsworthy and targeted toward specific key phrases.

DISTRIBUTION NETWORK

Your press release's success depends entirely upon the distribution network that the press release service uses. It will determine whether your press release is a small drop in the pond or whether it is able to create a ripple effect across the Internet. Before you put your press release in the hands of a press release distribution company, you need to ensure that it is in the right hands.

Be sure that your press release will be reaching Tier-1 newswires. The presence of "wire" in the name of the company does not necessarily mean that they function as a newswire service. Review the press release distribution service carefully and look for testimonials or recommendations that directly relate to journalism and news websites (i.e.: an association with CNN, Money Magazine or The New York Times).

There are three basic types of PR distribution that can all be accessed through one press release, if the press release is created correctly. First, push distribution utilizes emails and RSS feeds in order to deliver content directly into the hands of interested parties. For example, bloggers in the natural health field would subscribe to an RSS feed that will allow them to receive your press release on your vitamin company's case study, along with other relevant news.

If you opt for this route, you should be sure that your press release is being delivered in the body of the email or RSS feed and not as an attachment. Email attachments are less likely to be opened and can be flagged by some email providers as spam. Push distribution has the potential to reach an audience of tens of thousands. See if your press release distribution company offers email lists or RSS features that industry professionals can subscribe to.

Pull distribution has the same potential for reach, but it uses SEO optimization rather than direct delivery. Your press release is published on the press release distribution service's site and their page rank, along with your optimization, will increase your press release's chances of being listed in the search engine results for related terms. Your ranking will be determined by expert keyword usage and the traffic potential of your press release distribution service.

Finally, social media distribution is achieved by creating a press release that is interesting for the industry. It will encourage people to use social media and social bookmarking websites in order to take note of your press release content and share it with others. Case studies and industry reports are excellent fodder for stimulating this type of distribution. A press release distribution service should make it easy for people to bookmark and share your release if they are reading your press release on their site.

In addition to the three types of distribution methods that are utilized online, your press release distribution service should also give you the option of choosing a distribution area that will make the most impact for your business. For example, they should offer distribution in specific geographic locations. Not only does geographic distribution help laser target your press release, but it can also save you money. You shouldn't need to pay for national distribution if your news is related to your immediate geographic area. Options could include:

o International release – This category includes distribution to industry specific news sources worldwide.

o National release – This is distributed to national media outlets that are industry specific. This is perfect for press releases that deal with an industry as a whole.

o Regional release – There is normally a choice of one of four regions, Northeast, Southeast, Midwest and West/Southwest. This is a good choice for a local company that wants to expand to regional presence.

o State and local release – If your press release has a very narrow focus, you should limit the distribution to state and local release.

o Culturally specific release – Press release distribution services will have different names for these types of distribution. They are appropriate if you have news that is related to a specific cultural group (i.e.: African-American, Hispanic, Women). Your press release will be distributed to news outlets that are related to a specific cultural group.

In addition to regional areas, your press release distribution service should also let you choose the specific industry that relates to your news. In most cases, you can select five to ten industries that are relevant to the content of your press release. The more distribution options that your press release distribution company offers you, the better. Distribution channels help you match your press release for your audience so that it has the most impact. Remember, it doesn't matter how many people see your press release if they aren't the right type of people.

> *Case Study: PatioConnections.com*
>
> Distribution Network: OnlinePRNews offers four levels of service for their customers. PatioConnections.com selected the third level of service – the Multimedia Press Release Package. This package included submission to Associated Press and Top 100 Newspapers. Additional distribution channels for OnlinePRNews include print and broadcast media submission, submission to industry trade journals and submission to over 5,000 media websites (such as AOL.com, CNet News, Forbes and Bizjournals.com).

ANALYTICS

Analytics are the tools that a press release distribution service provides for you to track the reach and impact of your press release. A quality press release distribution company will offer real time, or close to real time, stats of your press release's distribution and impact.

For example, a press release distribution company could let you know how many times your release headline has been viewed, or what terms people are searching for when they find your release. At the absolute minimum your press release company should be able to give you links to your published press releases. To double check your reach, you should be able to find your press release listed on Google and Yahoo news search results pages.

> *Case Study: PatioConnections.com*
>
> Analytics: Online PR News offers a "Release Watch" report with their Maximum Media Visibility Package which offers links to actual releases as they appear on top and niche websites. PatioConnections.com did not choose to use this advanced service and relied on their own Google Analytics data to track results.

CUSTOMER SUPPORT

Customer support is essential when working with a press release distribution company. They should have a support desk that their customers can use and a 1-800 number listed on their website. Look for a company that offers training with creating a press release and clear guidelines for making your press release effective. Low quality services are only interested in getting your submission. There is less emphasis on making sure that the submission is going to be effective in garnering the attention of journalists, bloggers and other distribution outlet.

Most quality press release distribution services will assist their users with creating press releases by offering editing and evaluation services. In addition, they will confirm links and double check spelling before the press release is distributed to be sure that your company is putting its best foot forward.

Case Study: PatioConnections.com

Customer Support: Online PR News offers several points of contact for customers and potential clients. There is a contact form on their website, a login area for customers and a frequently asked questions section that helps inform future and current clients. They also offer editing services for their larger service packages.

RED FLAGS FOR PRESS RELEASE DISTRIBUTION SERVICES

+ **No guidelines for successful press release distribution:** A press release distribution company is your partner in distribution success. Guidelines for submission, tips for success and example press releases for you to model should be a part of any quality press release distribution site.

+ **No press release analytics:** You can only measure the success of your press release and plan your approach for the next press release if you have analytical data to work with. Your selected press release distribution site should offer comprehensive tracking capabilities so you know exactly what kind of impact your press release made.

+ **Limited distribution network or options:** Your press release's reach will be determined by the distribution network. The options you select will determine how your release is distributed.

Limited network and options will severely reduce the impact of your press release.

✦ **Lack of contact information:** The ability to reach a customer service can be critical in successfully using a press release distribution site. A company that does not have a clear line of communication for customer service does not have your interests in mind.

✦ **Offering free distribution services:** As you begin to evaluate press release distribution services, you will find that many offer free distribution services while some charge fees. Although there is some value found in distributing to free sites, there are a few major drawbacks that can impact the effectiveness of your press release. Free sites display advertising blocks in the layout of your press release, which can detract from the message. Free sites do not always display live backlinks, which limits the SEO benefits of distributing a press release. Free sites may also have a narrower distribution reach, which means that you'll need to send your same release to several different free sites. For the widest distribution and the lowest amount of hassles, using a paid press release service is your best bet.

CASE STUDY RESULTS

Although each individual press release will have results dependent upon the content and quality of the writing, looking at the results of our highlighted case study will help you realize the impact that a quality press release distribution company can have.

ONLINE PR NEWS CASE STUDY FOR PATIOCONNECTIONS.COM

Within 7 minutes of publication, the press release for PationConnections. com was appearing in the news results on page one of Google for several search terms. Shortly thereafter, the press release was distributed to Online PR News' strategic content partners and Tweeted on the OPN Twitter channel, resulting in even more saturation online. The following day, Dean Geyer wrote in to say that, according to his stats report, the press release published in Online PR News came was the 3rd highest generator of traffic and referrals to his site.

CONCLUSION

Online press release distribution can have a major impact to on your website's traffic and your business's standing in the market. The right distribution channels can help you achieve media saturation on a variety of different areas. Evaluating a press release distribution service based on their adherence to the previously outlined principles will help you determine whether or not they are worth your time and your money.

CHAPTER SIXTEEN:

SITE AUDIT SERVICES

A site audit should be an important part of any SEO company's work with your website. However, you can also purchase site audit services independent of other SEO services. A site audit will let you know what is going right and what is going wrong with your business website. A site audit service will help you make sure that your site is functioning properly so you can keep more visitors on your page.

In this chapter, we'll go over the basics of site audit services and what you should expect from a professional site audit company. Employing a site audit service will greatly increase your conversion rates because you'll be able to keep more of your customers.

THE IMPORTANCE OF SITE AUDIT SERVICES

Broken links, incorrectly rendered images, misspelled words and improperly formatted pages can all have a damaging effect on your credibility and make your site look unprofessional. Since the risk of fraud is so high on the Internet (because selling is not done face to face) it's vitally important that your website appear professional and put together.

Web site auditing services may help you catch areas that you and your web designers may have missed. Sometimes when you are so close to a project it can be hard to see mistakes that would be readily apparent to an outsider. In addition, a professional site audit company can point out areas of improvement that may not be readily apparent.

In general, site audit services will help you achieve:

o   A professional look

o   Lower abandonment rates

o   Removal of broken links

o   Increase load times

o   Reduce hard to reach pages

o   Improve META information that effects search engine results rankings

A site audit company will also look at a wide array of aspects that can affect your websites effectiveness include brand consistency, reputation, competition analysis, copy writing, conversion rations, design, site performance, on page and off page SEO, social media use and traffic analysis. In short, using a site audit company can help you improve your standing online and make sure that the traffic you are sending to your website is going to convert correctly.

Site audits are appropriate no matter where your site is in the development process. They are essential if your site is brand new and you are trying to make a splash in your market. If your site isn't giving you the quality of results that you are looking for, a site audit can help identify key areas of development. Site audits are also a good way to get a professional opinion without investing a lot of money into the SEO process.

Site audits can identify reasons why your site may not be appearing in the search engine rankings that you have expected it to. A site audit company can give you an overview of what needs to be fixed and then you can piecemeal together the fixes in house or step by step through various providers. To see a list of recommended site audit companies, see the rankings at topseos.com (http://www.topseos.com/rankings-of-best-site-audit-companies).

NEEDS ANALYSIS

The first step your site audit company will complete is conducting a needs analysis of your website. This will be conducted in order to understand the need for site audit services and identifying the purpose of your website. Until the site audit company knows what it is working with, it will be unable to analyze the data that is coming in and develop a plan for your site.

There are several different ways that your site audit company will analyze your needs once they've determined the purpose of your website. Your site audit company should ask questions about how you do business and how your website is used by your audience. They will determine where your traffic is coming from and how that traffic interacts with your content. They will pay close attention to the relevancy and accuracy of the information presented on your website. This is essential if your company is in a niche that is constantly evolving. Your site needs to stay current on important information relative to your business purpose. They should also look for additional ways you can present information, display advertising (if relevant to your business model), and increase your social networking presence by adding connections from your site to your profiles.

The site audit company should look at your main competitors and see how they stack up compared to the rankings and performance of your website. Analyzing the competitors may reveal some additional keywords and tactics that need to be used on your website. They should also point out ways that your site is succeeding when compared to your competitors so you know what site elements need to stay.

Ongoing analytics will be a major part of the site audit process. If you do not have an analytics program installed, this will be one of your site audit company's first recommendations. The analytics program will deeply detail the performance of on page factors and the incoming links from off page websites that improve your website's standing.

KEYWORDS ANALYSIS

Your site audit company will look at the keywords that your site is currently optimized for. They will compare those to your incoming traffic. In some cases, they may do their own independent keyword research and make recommendations for keyword optimization for your most important pages so they are more effective.

They will look into your current keyword rankings and determine if those ranking are helping or hurting the conversion of your traffic. You may find that you are getting a high number of visitors under particular keywords, but those visitors are leaving because your site has nothing to offer them.

The site audit company may identify entire niche sections that need to be addressed by your website. The keywords may have high search volume and high relevancy but aren't currently being represented on your site. The site audit company will recommend expanding into these areas with new site sections or blog posts.

In addition, they may suggest streamlining the existing content on your pages so each page relates directly to a particular keyword. If you've created a website in a "scatter shot" method and have added articles or pages just as you've seen fit, your site audit company will likely suggest that your streamline and focus your pages. Having the correct keywords on the correct pages will greatly increase your search engine traffic and relevancy for your users.

The site audit company's recommendation may result in more content and additional pages, but the results will be well worth it.

ON PAGE OPTIMIZATION

Once the site audit team has identified your keywords, they'll look at your website pages to see how well those keywords are reflected in your META data and your content. They will also identify links on your pages and determine if any links are broken. Having broken links on a page will create a frustrating experience for the user and it can lead to people leaving the site. The higher bounce rate you have, the lower your quality score and the lower your website will appear in the search engine rankings.

In addition to visible links, the site audit company should look at the hidden META data that is read by search engines and not users. There are opportunities to optimize your website for keyword terms in the META data, the images and the videos displayed on your website. The company should be able to identify ways that keywords can be used in the coding in order to increase your site's relevancy.

The site audit company should also analyze the server configuration and determine how your website deals with 404s, URL redirects and other technical matters. Analyzing and offering suggestions for these areas will help your site function more smoothly and it will lead to a better experience for the user.

A final area of on page optimization analysis will be how effective your website is at achieving your desired conversion action. The site audit company will analyze your bounce rates and see how users are interacting with your site. They may be able to identify key areas of your website that are turning your audience away and help you identify ways that you can improve your conversion rates through information architecture, design or web copy.

OFF PAGE OPTIMIZATION

Off page optimization analysis techniques are important so your company can develop a method for building backlinks and increasing your relevance within the niche. Better backlinks lead to better search engine results. For more on how to build quality backlinks, see the link building chapter of this book.

Your site audit company will do a comprehensive review of the incoming backlinks and related keywords. They will identify any major problems with your current backlinking strategy and give you suggestions for how you can increase your backlinks. They will let you know which keyword terms and niche contexts are currently the most prominent in your existing backlinks. They will assist you with determining if the existing backlinks meet your keyword needs.

They will also analyze the quality of those backlinks. As mentioned in the link building chapter, not all backlinks are created equally. If you have a high number of backlinks coming from low quality sites or spam sites, it's not doing your site any favors. They will suggest methods for improving the quality of backlinks and may pull in examples from a competitor's site to show which third party sites need to be approached for backlink building.

REPORTING METHODS

Generally, a site audit company will deliver your report in a PDF document that details their findings and recommendations for implementation. This document is essential because it will provide your company with a firm foundation for the steps you will take in the future to make your site more optimized.

A site audit company, if they also provide SEO services, may also deliver you with a price approximation for fixing the problems on your website. This should not be an obligation to purchase services from them but it can be helpful as you start to price for services for optimization.

When you hire the site audit company, they should be clear about how the information will be delivered to you. They should also be clear about the level of detail of the information that you will receive. The word "report" has many connotations and you can only be sure of the quality of the deliverable if they are specific on what categories they will cover.

RED FLAGS FOR SITE AUDIT

- **No needs analysis step:** While there are best practices that your site audit will compare your site to, there are also considerations that should be made on a case by case basis. Before getting started, your site audit company should analyze your company and website's purpose and view the best practices through that lens in order to provide a customized evaluation.

- **No keyword suggestions:** A professional site audit company will offer keyword suggestions and not simply look at the keywords that are currently being used on your website and inbound links. They will look for opportunities to increase your optimized keywords both on the page and through backlinks. If their site audit analysis does not include keyword suggestions, look elsewhere.

- **Limited on page or off page analysis:** Both on page and off page optimization are important to having an effective website. If your site audit company doesn't offer a balanced evaluation of your website with both on page and off page factors, they aren't providing you a complete evaluation. Be sure to ask extensively about their evaluation methods and review a sample report to make sure both areas are covered.

- **Incomplete results report:** Seeing a sample results report will determine whether or not the site audit company is capable of giving you the quality of information you need to move forward with your site's optimization.

+ **Insisting on hiring the company for other SEO services:** Site audit services should be an independent service that you can purchase apart from other SEO services. Hiring a company for site audit services is not making a promise that you will hire them in the future. Free site audits as part of a larger package of SEO services are okay to purchase so long as you evaluate the SEO company underneath the criteria outlined in other chapters of this book.

CONCLUSION

When evaluating site audit company's services, look at the value of their site audit and not the value of the other services that they offer. The site audit services should be evaluated independently of anything the company may or may not offer. Look at the previously outlined criteria and inquire about previous customers. Follow up with these customers to make sure the site audit company did a good job and offered real advice that was useful. Finding a reliable site audit company can be the foundation for successful website optimization.

CHAPTER SEVENTEEN:

SEO TRAINING PROGRAMS

S EO training programs can help you implement any of the principles in this book for your own company. For some SEO areas, you may find that your specific knowledge about your industry, combined with training from a reputable program, will be enough to help you make your mark online.

When you begin searching for an Internet marketing training program, it's important to do your due diligence. Training programs can often be nothing more than make money online schemes in disguise. Seek out a program that specializes in helping businesses like yours establish and grow an effective website. Avoid programs that emphasize making money quickly or using "black hat" methods to get traffic and rankings. There are hundreds of programs out there that are only designed to make their creators money and don't have any relevance to real SEO.

CHARACTERISTICS OF VALUABLE SEO TRAINING PROGRAMS

Using the criteria outlined in this chapter you'll be able to weed out the useless programs from the ones that are produced by real industry leaders who have a track record of delivering helpful, useful information. Before you select a training course, review these evaluation criteria and apply them to the area of website training or SEO that you most need to be able to reach your goals with your website. You can also view a list of top SEO training programs at http://www.topseos.com/rankings-of-best-training-programs-companies.

COMPREHENSIVE TRAINING

A reputable Internet marketing training company will either provide a variety of training programs or offer one comprehensive resource that covers a variety of topics. As you've read throughout this book, there are many areas of Internet marketing that are required in order to make a company's website successful. It makes sense to locate a training program that covers a wide variety of topics.

You will likely find that reputable training programs will have an introductory "overview" course that will cover many different aspects of effective web management – from keyword selection to on page SEO to PPC advertising to landing page optimization. The company may then offer specialized courses that are designed to educate you with details on the various aspects of website optimization and management. Finding a company that offers both the broad overview courses and more detailed classes is preferable because you can go more in depth on topics that are relevant to your stage of business.

The introductory class should contain coverage on all aspects of Internet marketing that are relevant to your business. Review the sections of this book to understand how these different components work together and then seek out a training program that contains training on all of these areas.

VARIANCE

Business owners who want to use Internet marketing principles to promote their businesses come in a variety of different skill levels and with a variety of different goals. Seek out a training program that offers different levels of training that are customized to where your business is at right now. For example, some training companies offer beginner, intermediate and advanced courses. Others break down training into specific sections, ex: Internet marketing for non-profit organizations.

Another type of variance is the usage of different methods of training. Some SEO training companies concentrate on doing live seminars while others specialize in home study courses via video. There are benefits and drawbacks to each method of training.

o   Live seminars: The pro of this type of training is being able to network with other business professionals who need to increase their Internet marketing knowledge. You can also ask questions of the presenter and get a large amount of information in one or two days. There is a one-time cost for live seminars. The downsides are that you may not be able to ask follow up questions and may forget some of the information over time.

o   Membership site training: With an SEO training membership, you'll pay a monthly fee to get access to online training. There are normally training modules delivered via text, video or audio. There may possibly be a membership forum that you can use in order to collaborate with others and ask questions. If you go this route, you may have higher costs overall than a one-time event, but you'll get the ongoing support you need.

o   Fixed term class: A fixed term class can be done in a live event format or online. A fixed term class offers ongoing support for a fixed period of time and also more in depth information than can be gathered at a one-time event. Since there is a defined structure to the class, you're more likely to get the information that you need quickly, as opposed to a membership site training format where you may need to be subscribed for months on end. Fixed term classes are also more likely to be focused to a specific aspect of Internet marketing.

An Internet marketing training company may also offer a variation of these types of trainings, presenting some information in a live format and some information in a membership site model. Thoroughly investigate the different options that the Internet marketing company offers. You may be able to get the same information in a format that appeals to your learning style or your company's goals.

CLEAR TRAINING

There is no point in obtaining Internet marketing training if you can't understand how to implement the information. The training you receive should be clear and easy to understand. Finding a training program that is geared toward your skill level and your industry will be essential in helping you understand the training.

The training program, event or course should have a clear goal in mind. The Internet marketing training company should be able to define the goals for the students and what your expectations should be of the training. Look for specific goals and guidelines rather than over the top claims for instantly tripling your traffic over night, or other outrageous claims. Reputable companies will take a teaching approach to the curriculum, rather than hype up their achievements.

The training should have actionable steps that you can utilize after the training is over. It's one thing to learn about the concepts involved in SEO and how they can help your company, and it's something else entirely to be able to translate that training into a plan for your business. Compare the difference between learning about how pay per click marketing works and learning how to research keywords, write ads and bid on keywords. The latter will produce a much more thorough and actionable understanding of the concepts of PPC that can be used by a company to get real results.

Make sure that the Internet marketing training that you are going to be participating in is designed for business owners and not industry professionals. There are many SEO training programs and courses that are aimed at helping SEO service providers assist their clients with better search engine rankings. These courses may be too advanced for the average business owner who has the goal of increasing traffic and conversions for their specific website.

A program that is geared for business owners will be delivered at a level that is understandable. It will give you the information that you need to achieve Internet marketing success without going too in depth on the concepts. Remember, you don't need to learn all of the ins and outs of SEO. You just want to get the specific techniques down in order to be able to see results on your website.

See if you can obtain a preview of the training delivered by the program to gauge the quality and the comprehensiveness. If the information in the preview is delivered in a clear and concise manner, you should be able to understand and implement the information presented in the full course.

Previews may not always be available, but they may have articles or blog posts on their website which will give you a sense of the quality and clarity of the paid training.

Also, research the names of the trainers, founders or teachers. Most will maintain an independent blog or website, or have published work under their name. You can get a sense of their teaching style, their expertise and their level of clarity from their other published work.

UP TO DATE TRAINING

Search engine optimization changes rapidly and your Internet marketing training needs to be up to date in order to have an impact on your business. Make yourself familiar with some recent SEO and SEM developments, and then ask if your training program will cover those developments.

Some resources for keeping tabs on the industry are:

o   Search Engine Watch – http://www.searchenginewatch.com (SEO and SEM)

o   Mashable – http://www.mashable.com (SMO)

o   SEO Pros – http://www.seopros.org (SEO and PPC)

Another way to determine whether or not a training program is staying up to date is to read articles and blog posts on their website. If it seems like they are staying up to date on industry news and are posting content that is fresh and relevant you can be assured that the training program will be kept up to date as well.

This does not mean that a brand new company will be better at training than a company that has been around for a while. Established Internet marketing training companies make it a habit of updating their information and staying on top of the industry trends.

FEEDBACK FROM CLIENTS

Reading the feedback from previous clients will be a good indicator of how the company has performed in the past. You can read testimonials on the training company's website, but it's a better idea to search online to see what the general public thinks of the training information. If you have forums or web groups dedicated to your industry, you can also ask your peers about Internet marketing training. You may find a company that deals specifically with your industry that others have worked with in the past.

The Better Business Bureau and consumer feedback are excellent sources of determining the trustworthiness of an Internet marketing training company. By thoroughly researching a company before you pay for the services, you can ensure that you'll be working with a reliable company who can give you the information that you need.

RED FLAGS FOR TRAINING PROGRAMS

- **No free information on their website:** Steer clear of a company that has a website that consists of just a high pressure sales letter. Reputable training companies will freely give helpful information on their websites in the form of articles and blog posts. By reading this free information, you can get a sense of what you will be learning from them.

- **No guarantees:** Although guarantees for SEO or PPC services should not be trusted, training should have a guarantee. They should guarantee that the information you'll receive will be actionable and up to date. While they cannot promise that you'll see results with their methods, they should promise that the information that you receive will be valuable.

- **One level of training:** Reputable Internet marketing training programs are designed to help people at all different levels and as such, they normally have several levels of training to best meet the needs of their students. It is unlikely that a basic class will be able to give you everything you need for the rest of your website's existence. Having the option to go back and learn intermediate and advanced training from the same company is essential.

- **No feedback or negative feedback:** This is a given. You shouldn't trust your Internet marketing training with a company that does not have feedback from previous customers.

CONCLUSION

Investing in SEO training for your company is a wise move. Not only will it equip you with the skills to make changes to your website for optimization purposes but it will also help you determine the quality of SEO providers if you need to employ their services in the future.

CHAPTER EIGHTEEN:

WEB ANALYTICS SOFTWARE

Web analytics software gives your company the accurate data that it needs to determine the traffic coming into your website and that traffic's response to the website. A good web analytics tool will track important data and understand what is happening on your site so you can make effective plan for your future development.

Web analytics tools can answer the following questions for you:

o   How many people visit your website each day?

o   What are their initial impressions of your website?

o   What do visitors do when they arrive at the website?

o   What are the most popular features or pages of the website?

o   What areas of the site need to be improved upon?

Armed with your site's statistics, you'll be able to understand your visitor's needs more effectively and optimize your page so you can convert more visitors into buyers. Web analytics tools can automate the process of comparing traffic statistics to your sales. Web analytics tools can also give you more in depth information than what you could gather on your own. In short, you'll be able to gather more data in a shorter period of time so you can laser target your website.

The following chapter will go over several qualities that you should look for in a web analytics program. For more detailed recommendations, visit http://www.topseos.com/rankings-of-best-web-analytics-software.

BASIC WEB ANALYTICS FEATURES

Many web analytics programs share a variety of features that you should look for. Keep the following features in mind while looking for a web analytics tool. They should be included in the tracking dynamic for your web analytics program.

o *Visits* – This is the most useful area of tracking because it tells you how many people are coming to your website and visiting your pages. An overall number of visits to your site should be increasing over time to get an increase in results. By comparing the number of visits received on each page you can see which of your pages are the most popular.

o *Unique visitors* – Not all visits to your site are created equally. If your web analytics program is just looking at the number of visits and not the number of unique visits, it is doing your analysis a disservice. If two people visit three times a week, only their first visit should be counted.

o *Page views* – This will tell you how often pages on your website are visited. The page view count can give you valuable data on the most popular parts of your site. You can use this data to maximize the exposure of this particular page or retrofit the other pages to match the popular ones in design and content.

o *Top entry and top exit pages* – These pages represent the top places where the most visitors enter and exit your website, just like the term sounds. It will give you insight into what pages are gathering traffic from search engines and which pages are causing your visitors to leave.

o *Referrers* – This metric shows the external links that are coming into your website and referring traffic your way. This can be useful if you notice a big flow of traffic into your website. It will help you identify where the traffic is coming from.

o *Search keywords* – The search keywords are the words that are used by people to find your website. This can be valuable information in developing additional content and sections of your website.

o *Visitor information* – An analytics tool can tell you what country your visitors are coming from and what region of that country they are coming from. It can also tell you the browser type and size they are using in order to view your site.

o *Click paths* – Click paths are also called click tracks and they tell you how your visitors journey through your site. Advanced analytics program will be able to show you an overlay of your client's movements on the page so you can identify the most popular areas of your website and whether your conversion areas are getting the attention that they should be.

o *Conversion* – This statistic is a bit more involved but should be part of a more advanced web analytics tool. Conversion statistics will track the number of people who take the action that you want them to take on the page. For example, it will determine how many people filled out a form or how many people purchased your product.

o *Statistical overlay* – A statistical overlay will show you the page hits, session length and unique visitors that come to your site in a graphical format. The data will be displayed over your webpage so you can see exactly what is going on with your visitors.

o *Market segmentation* – The visitors to your website can be divided into a variety of different categories. Frequent visitors, infrequent visitors and visitors from a variety of different locations may interact with your site in a variety of different ways. Your web analytics program should allow you to segment out your traffic into different categories so it can be better analyzed. Your analytics tool should track keyword buys, email campaigns, banner ads and your other forms of marketing.

o   *Sales funnel follow through* – As a visitor arrives on your site, they'll go through a series of steps in order to get to your conversion action. A good web analytics program should be able to track these steps and give you an idea of how the customers are following through.

o   *Form abandonment* – Any form on your site is an opportunity for your visitor to take another step with your conversion action or to walk away. Tracking the rate of form abandonment can help you a great deal in understanding how your site needs to be optimized.

Advanced analytics tools may also include metrics above and beyond what it listed here. However, this should be the bare minimum of what they offer in their web analytics tools. Analyze the tracking metrics that the web analytics company uses before you invest in a web analytics program. The following areas should also be paid close attention to while you are choosing a web analytics program.

IMPLEMENTATION

Implementation is the ability to integrate your analytics program into your existing website. You should search for a program that can be easily installed into your website's framework so it can track your traffic and your visitor clicks. The web analytics program should either be accessible enough that you can install yourself or you should be able to get assistance from the company you're purchasing the analytics program from. Look for a web analytics program that is built to integrate into your existing website type of framework. For example, many web analytics programs have an easy to use widget that can make it simple to implement the program into a WordPress site.

USER INTERFACE

If you can use a trial version of the web analytics program before you purchase it, you'll be able to tell if the user interface will meet your needs. Look for an interface that is intuitive, displays all the important information in one place and that allows you to customize your reporting options. Much of the user interface qualities will be dependent upon your personal preference, so it's pertinent that you find a web analytics program that displays screenshots or allows you to try out their service before you buy. Getting a sneak preview will let you know if you'll be able to analyze the data effectively and get use out of the software.

TIMELINESS

This refers to the amount of time it takes for an action to happen on your website and then appear in your analytics. Always opt for real time web statistics. Real time statistics will update automatically and provide the most up to date review of your website's activities. Real time data will allow you to make changes to your traffic as they occur. For example, if you have a sudden spike in traffic you can immediately see where that traffic is coming from and make necessary changes to your site. If you have an increase in traffic due to being mentioned on a popular blog, you can change your home page to welcome visitors from that page.

ACCURACY

No web analytics are 100% accurate. It's a fact of the industry and your analytics program shouldn't claim that their results are 100% accurate. In the best case scenario, the web analytics that your program delivers are a best guess. However, they should be precise. Precision is the ability to repeatedly product certain types of results. Your web analytics program should give you precise data that can be acted upon. For example, if the data tells you that you've had 10,000 visits per week, it doesn't mean anything but if it's a 10,000 jump since the week before, that represents data that you can work with. You should look for a web analytics program that can offer you consistent data and an exact method for collecting that data.

CUSTOMIZATION

Web analytics software should offer you a deep level of customization to create customized reports and tracking metrics that are in tune with your business model and your needs. Advanced customization features can include advanced segmentation, custom reports and a customized dashboard. Custom variables will allow you to define multiple tracking segments that are based on a variety of factors. These will result in unique site data that will help you customize the web analytics data to meet the needs of your business.

RED FLAGS FOR WEB ANALYTICS SOFTWARE

- **Analytics programs that offer too little customization:** Getting customized reports from your analytics software will make a big difference in how you're able to analyze and use the data for optimizing your site. Although most website owners will have some basic commonalities in their use of analytics, the program should be customized for your own purposes.

- **Promises of 100% accuracy:** Accuracy is impossible at a 100% rate. Your web analytics company should be focused on creating precise results that are consistent from month to month in their data collection.

- **Incomplete list of features:** Free tracking programs and "lite" versions of analytics programs can get you started but in order to get the most out of your business, you should invest in a program that will meet all of your long term needs.

- **Clunky user interface:** You're less likely to use a web analytics program if the interface doesn't make sense to you or is too complicated to understand. It should streamline the process of analyzing data rather than slowing you down.

CONCLUSION

Web analytics can help you make the most of your website and plan for your future development. If you're not currently using some type of analytics program it can severely limit the ability for any of the SEO providers listed in this book to help you achieve results with your site. Use the previously outline criteria in this chapter to evaluate and select a web analytics program that will work right for you.

CHAPTER NINETEEN

EMAIL MARKETING SOFTWARE

Email marketing is a reliable form of communicating with your audience. Sending email newsletters and marketing messages ensures that you will be able to capture your visitor's important information while they are on your site, which means that you can market to them in the future and increase your conversion rates. With a reliable email marketing software program you can give your business a lasting presence online.

Email marketing software generally includes the tools to capture email information from visitors and assistance with organizing those visitors into a database. You'll also have the capabilities of delivering email messages to your database at a pre-determined time or automatically upon the users' request. Email marketing software can be used on your company's servers or you can utilize a web-based email software program. No matter what your choice is, you need to evaluate the software for reliability and compliance with spam standards.

In this chapter, we'll look over various methods of evaluating a quality email marketing software program. We'll discuss why email marketing is so important to the success of your web marketing efforts and how finding the right marketing software can make a big difference in your results. The following section contains several areas which you can use to evaluate an email marketing program.

THE IMPORTANCE OF EMAIL MARKETING

Email marketing utilizes email to deliver information about promotions, company offerings, product updates and (most importantly) helpful information about the topic of your website. While email marketing originated as a form of "digital direct mail" the trend has changed away from direct sales strategies toward building a relationship with your market by offering helpful information and other forms of connection (i.e.: a notification about an update on your company's blog).

Email marketing is cost effective and delivers better results than traditional direct mail marketing. If you're doing business exclusively online, this may not seem like much of a problem. But for those who have tried traditional mail marketing and other expensive forms of getting the world out, email marketing can seem like a fresh breath of air.

People visit dozens of websites per day and may even visit hundreds if they are on the quest for important information, reviewing a product or are just Internet-inclined. If they come across your website and don't find what they need, or don't understand what your site is about, it is almost guaranteed that they will not come back. Even if they take interest in your product or service and bookmark your site, chances are they will rarely come back to visit your site and take part in your sales funnel process.

By offering a free report, email course, discount, training or newsletter in exchange for an email address, you can capture the lead in the moment and bring them back into your sales funnel long after they've left your site. You'll get more repeat traffic, have an opportunity to build your brand with your customer base and give value to them in ways you just can't with a static website.

In addition to collecting email addresses and distributing messages, your email marketing software program will also help you manage and automate email marketing. A comprehensive program will allow you to test message subject lines, see open rates, segment your marketing list depending on what actions they've taken and analyze click through rates. In short, a helpful program will help you see how your list is growing, who is taking what action and how you can approach those action takers more effectively.

The following qualities should be analyzed while you are searching for the right email marketing software solution for your company's needs. While your needs will be unique to your company and your use of email marketing, there are some basic practices that should be adhered to in order to make sure that your money is being spent wisely in this area.

SUBMISSION INTERFACE

Depending on the nature of the email marketing program, you may be interacting with the submission interface on your own server or on the email marketing company's website. The submission interface will include creating a message (including the subject and body), selecting the portion of the list to deliver the message to and setting up a time for delivery (whether it will immediate or sometime in the future).

Most email marketing software programs offer two basic types of messages – follow up messages or broadcast messages. Follow up messages will automatically be sent to list members as they join your list. They are set up to be delivered in a specific sequence and will go out to anyone who becomes part of the list. You can use the follow up message feature to set up a free course for your list members. They will receive each message in sequence no matter when they join the list.

Broadcast messages are useful for monthly newsletters, special announcements and other marketing messages that are timely in nature. They are created in much the same way that you'll create follow up messages but you'll arrange for their delivery either immediately or at a specific date and time. At the arranged delivery time, all members of your list will receive the message at once.

The messages aren't the only items you'll be creating in the email marketing system. You'll also need the capability to create web forms that will capture email addresses on your website. It should be simple for you to add and remove information areas on the form. For example, you may want to capture the name, email address and location rather than just the name and email address. You may want to use a more detailed web form for specific marketing activities. Look for a program that offers fill in boxes, drop down menus and radio buttons for information gathering purposes. Once you generate the web form, you should be able to integrate it seamlessly into your website using HTML or JavaScript.

In addition to the separate types of delivery methods and web form generators, an email marketing software program should be accessible from any type of browser, and under a variety of languages if that applies to your business needs. Most professional programs will include pre-configured

367

action forms and templates that you can use in order to get up and running with your email address collection process.

HTML templates can create email marketing messages with more impact. The messages will be more memorable to your list members. Considering how many email messages the average person gets per day, any steps that you can take to make your messages stand out should be sought out. Your email marketing program submission interface should allow you to easily integrate HTML formatting and design into your email messages. Integrating your company's logo through an HTML template should be simple as well.

Look for an email marketing program that has many "wizards" to help simplify the message generation, lead capture and email formatting processes. Wizards are step by step interfaces that guide you through the entire process of a particular step. They help reduce the learning curve so you can work with your email marketing program much more quickly. The more quickly you can get your forms and messages in place, the more quickly you'll be able to build your list and market to your niche.

PROTOCOL COMPLIANCE

It is easy to get started with email marketing, but the marketing form is not without its limitations. The CAN-SPAM Act of 2003 established some strict rules for email marketing in the United States which all online business people need to be aware of. A similar law was passed in the European Union the year before and the UK passed the Privacy and Electronic Communications Regulations in 2003. An email marketing program should assist you with meeting the rules for email marketing.

In the case of the CAN-SPAM act, there are stiff penalties for abusing email addresses with sales messages that aren't requested by the user. There's a $11,000 penalty per violation for sending spam (unwanted email messages) to an individual user. It's clear that using an email marketing system that works within the CAN-SPAM protocols is essential. The law was updated in July of 2008.

The basic rules for CAN-SPAM compliance, and its European counterparts, are

o   The subject line cannot contain misleading information.

o   The subject line needs to clearly identify advertising if it is present within the email message.

o   Email headers, "from" addresses and other identifiers cannot be tampered with in order to conceal identity.

o   The body of the email must contain a valid physical address for the sender.

o   The email must contain a functioning opt-out mechanism.

o   Opt out requests must be honored within 10 business days of receipt of the request.

(FTC.gov)

An effective email marketing company will help you avoid CAN-SPAM violations by setting up a few safeguards so that your messages will be delivered safely. Most email marketing companies will automatically include an unsubscribe link within your email marketing messages. They will also require you to enter a valid real world mailing address while setting up your account which will be included in your email message.

Another frequently used safeguard feature of email marketing programs is to rate the messages based on criteria to determine their likelihood of being caught in "spam" filters (which most email systems use). Including words like "free", "make money" and even "spam" will rate a message as more likely to be spam. Your email marketing message should give you notification if your message is verging on being seen as spam.

Your email marketing software should use double opt in confirmation in order to ensure that the messages are welcomed by the recipients. Double opt in confirmation requires all new list members to verify that they've requested to receive information from you before the first message is sent. This will reduce the likelihood of your messages being marked as "spam" by the receiver. If enough receivers mark your messages as "spam" your email address will be prevented from sending email to that particular email client. You can ensure that your messages are welcome and treated correctly by the recipients by utilizing the double opt in capabilities of your email marketing software program.

DELIVERY RATE

An email marketing software program should ensure that the vast majority of your email messages that are sent are received by your list members. A lot of deliverability issues can be addressed with double opt in confirmation, as discussed earlier. This will ensure that your messages are welcomed by your list members.

In addition, your email marketing software provider should authenticate email campaigns through a series of processes. First, they should set up Sender Policy Framework (SPF) records on your server if you are hosting the program or on their own if you are using a web based email marketing program. SPF records will let the receiver's host verify that the email is being sent from the location that it claims it is being sent from.

Your mail host should also enable reverse DNS on your server, or theirs, depending on the set up. Reverse DNS is another safeguard that will enable the email provider to verify that the email is being sent from a real sender. It will allow the email provider to look up the domain name and host that is associated with the email being sent. If the reverse DNS lookup request results in "no domain associated" then the email provider will reject the email. By utilizing reverse DNS lookup, your email marketing software program will assure that your email messages will be delivered no matter what email provider the receiver is using.

DomainKeys is another email authentication system that your email marketing software company should utilize. The DomainKeys authentication system will notify the receiver's email provider that the email is coming from the domain that it claims to be coming from. DomainKeys should be installed on your server if you are using a server based email marketing program. If you are using a web based service you should verify that they are registered with DomainKeys. DomainKeys is used by popular email clients Yahoo and Gmail alike.

Your email marketing software company will likely have established relationships with the major ISPs. Look for evidence of this on their features list. They should make it clear that they successfully deliver emails to the most popular ISPs. Even the casual Internet user should be able to identify the most popular ISPs. Look through the list of providers and if you see any major ones missing, look elsewhere for your email marketing needs.

Deliverability rates can be significantly harmed if unsubscriptions and undeliverable addresses are kept as part of your contact list. Your email marketing software program should automatically filter these addresses out as they become unusable. This will help protect your website and email address from being labeled as spam.

A professional email marketing company will stay on top of new developments in SPAM laws so that your marketing campaigns can stay compliant. Look for a company that maintains an active blog and keeps its customers abreast of new developments in this area. Having new techniques, filters and checkpoints for email deliverability is a good sign. It proves that your email marketing company is staying up on new laws and changing their protocol so all of their customers can be successful.

ANALYTICS

Understanding where your emails are being delivered, how many members of your list are opening them and what actions they are taking is important in ensuring that you can use email marketing effectively. With analytics incorporated into your email marketing program, you can better understand your campaigns and make improvements in them in the future.

Your email marketing company should provide you with a variety of different analytics devices that will tell you about your campaigns so you can make improvements in the future. The email marketing program should have at a minimum an analysis of open rates, click rates, unsubscriptions and bounces. Added features could include statistics to see sales revenue that is generated by your opens. You should also be able to segment your list based on their activities so that list members that take a specific conversion action are transferred to another list so you can further your marketing activities with that specific group.

You should look for a company that gives you the ability to split test email subject lines and body messages so you can test for effectiveness and select what works best for your particular list. The company should provide real time statistics in a variety of different means (through graphs and lists) so you can access the data and act on it.

SUPPORT

Support is essential with an email marketing program because it will help you send marketing messages successfully that actually get results. Support can come in many different varieties, but for the best results with an email marketing company you should look for a company that provides as many of these qualities as possible.

Support traditionally comes in the form of a help desk, live chat support or a customer service number. These resources will be helpful if you encounter problems with using the email marketing program that require immediate attention. Support should be offered on a 24/7 basis so that you can get the support that you need when you need it. You may also find support in the form of an FAQ database or support forum, but this form of support should be in addition to the live support and not a replacement for it.

Additional support that should be sought out is support in the form of tutorials, videos, articles or reports that assist you with setting up your email marketing campaigns. The help should show you the logistics of using the program but also give you tips for effective email marketing. You can get a good idea of the amount of support that a company provides its customers by doing some research in marketing forums and looking through reviews of email marketing providers.

RED FLAGS FOR EMAIL MARKETING SOFTWARE

+ **Free email marketing messages with advertisements displayed:** Nothing is more unprofessional than sending out an email marketing message with an advertisement for the autoresponder program on the bottom. Invest in a subscription based email marketing program that is advertisement free.

+ **Limited email marketing message types:** An email service should allow you to create both follow up and broadcast messages, as well as plain text and HTML messages. Look for various HTML templates available for you to customize so you can give your email marketing messages a professional look.

+ **No web form wizards or web form customization:** In order to start building your list, you need to gather email addresses. Your email marketing list should make it easy for you to create a form that will collect email addresses, names and other important information from your website visitors. Creating a web form in the format that you need is essential to developing an email marketing list.

+ **No evidence of protocol compliance:** A successful autoresponder company will deliver messages within the guidelines of the CAN-SPAM laws. They should make it clear that their company works with the basic protocols of effective email delivery so that you can be sure your messages are arriving where and when they should be.

+ **No safeguards against SPAM:** Your email marketing company should make you aware of up to date SPAM information and include safeguards within the system to prevent you from possibly sending messages that could be construed as spam.

+ **No assurances of delivery rate:** Any professional email marketing company will take pride in their high delivery rate. Look for a company that boasts a near 100% delivery rate. The closer the rate is the better.

+ **Limited analytics capabilities:** In order to use email marketing effectively, you need to be able to review and adapt your marketing based on your previous results. Avoid a company that offers limited analytics because it will not help your success in the long run. Review a sample report of what they can offer and study their analytics features before investing.

+ **No split testing capabilities:** Split testing is the best way for you to test the effectiveness of subject lines and other conversion factors within your email list. Avoid an email marketing program that does not include split testing capabilities.

+ **Insufficient support for your needs:** The level of technical support you may need with an email marketing program that will be entirely dependent upon your level of technical skill. If you or your team is brand new to email marketing and need step by step guidance, look for a company that will offer you hands on support. Alternatively, if you are comfortable with technical issues, hands on support may be less of a concern.

CONCLUSION

Email marketing software is a part of the foundation of your online business so it deserves attention and consideration. Switching email marketing software programs after you've started collecting email addresses will be difficult, so it will pay off in the long run to invest time now in carefully evaluating the providers and their offerings. Try to obtain a free trial period with the provider so you can test out their system before you buy. Using the criteria listed above, you'll be able to correctly evaluate a provider for their effectiveness and professionalism.

CHAPTER TWENTY:

MARKETING AUTOMATION SOFTWARE

Gathering leads from your website is important to building a long term, sustainable business. There are many aspects of online marketing lead collection that can be automated in order to free up your time so you can focus on other aspects of your business. With the use of marketing automation software, you can collect and organize customer data in a way that will increase your effectiveness.

Marketing automation software can streamline your lead gathering and scoring process so you can focus your attention on nurturing those leads toward a sale. This type of software allows your company to focus on the activities that will make you the most money.

Marketing automation software can automate several marketing processes like:

o Customer segmentation

o Customer data integration

o Campaign management

o Customer scoring

o   Visitor Tracking

A fully featured marketing automation software program can make the difference in whether or not you'll be able to gather information from your leads and move them successfully through your sales funnel. In this chapter, you'll learn about the capabilities that a marketing automation software program should include in order to be helpful in streamlining your business.

Any standard email marketing system or analytics program will tell you where your customers are coming from and what their contact information is. However, a fully featured customer list won't give you the in depth information that you need in order to build opportunities with your customers. Marketing automation software will help you create cross-selling and up-selling situations that you may not be aware of otherwise.

MARKETING AUTOMATION SOFTWARE QUALITIES

The tasks that marketing automation software completes can technically be done manually, but you won't get as precise results as quickly as you do with software in place. The investment in marketing automation software is an investment in the growth and management of your business. However, you shouldn't jump into the decision of buying marketing automation software without making sure that it will serve your needs. The following qualities should be sought out in a marketing automation software program. In addition to these qualities, you can see a current list of recommended marketing automation software programs from topseos.com at http://www.topseos.com/rankings-of-best-marketing-automation-software.

Case Study: Rare Space

Marketing Automation Software: Net-Results (http://www.net-results.com/)

Rare Space is a Denver-based tenant advisory firm that specializes in commercial property tenant/buyer representation. The commercial real estate industry presents a challenge because the buying process is lengthy, expensive, and only executed once every few years. Rare Space's ideal lead is a C-level executive (CEO, president or controller) but contacting them online proved to be quite difficult even though Rare Space was at the forefront of standard email marketing.

Rare Space had been utilizing standard email marketing practices and tools for about five years but lacked crucial intelligence on where they were in the buying process. Once their recipients opened their emails, Rare Space could no longer track the behavior of their prospects, had no way to determine what their prospects were interested in, and where they were in terms of their leased or purchased building.

Case Study: Rare Space

Ease of Use and Implementation Time: Net-Results was able to offer Rare Space the ability to build upon their existing leads list as well as build a foundation for future growth. They were easily able to identify which portions of their website their prospects were visiting.

EASE OF USE

The easier marketing automation software is to use, the more likely your company will be able to utilize it efficiently. The marketing automation software should be completely customizable to your company's needs. Look for a software program that a variety of different features that can be "turned on" and "turned off" according to your company's needs. All of the important data should be collected and displayed in one centralized location so you can see at a glance how your data is being collected and handled.

Another feature to look for in marketing automation software is tutorials and technical help for new users. A comprehensive marketing automation software program will include video or written tutorials to help your company hit the ground running and use the software quickly. There's no point in investing in software that is not usable or actionable immediately. You're making a purchase to solve a problem and not to give yourself or your team more tasks to deal with.

Once the program is up and running it should aggregate your data, label your contacts and deliver results without much input on your end. Lead cultivation and lead qualification should be done automatically, with the option of being able to handle the tasks manually as well. The results should be in real time so that the appropriate sales person can respond in a timely manner. Reports should give comprehensive information about your site's visitors and their interactions with your website so you know what steps need to be taken next. The lead scoring system should be easy to understand and the program should allow you to easily take the next marketing action with that lead based on their interaction with your site and the quality of the lead.

IMPLEMENTATION TIME

Implementation time refers to the amount of time that it takes for your company to implement the software into your website and your marketing process. If you spend too much time trying to accomplish integration, it will prevent you from being able to use the program effectively.

Look for a software program that will integrate into your current marketing framework and that is customizable to your needs. Most programs promise implementation within just a few days of purchase, which can get you up and running with the system quickly. Apart from initial implementation time, you should also look for evidence of how quickly you can set up the various tracking elements and coordinate the lead generation with the system you are currently using. Basically, read the details of how your possible program tackles integration and search for reviews from customers to see the reality of their promises. Most companies will claim that integration is easy and simple, but only customers will know how those promises translate into reality.

Case Study: Rare Space

API and Integration: Rare Space was able to link their existing database with Net-Results. This saved on time and allowed them to make better use of the existing leads. Using Net-Results re-energized their database and helped them make use of "middle of the funnel leads." Net-Results integrates with VerticalResponse, Salesforce.com, LinkedIn, Twitter and Jigsaw.

API

An application program interface (API) is used by a software program in order to integrate with other software programs. Likely your company is aggregating leads from various other programs, like email marketing systems and social networking sites. An API will allow your marketing automation software to bring in information from those systems into the software program for further use.

Look at the list of the programs your company is currently using to connect with your target market and compare that to the features of the program you are considering. It's important that your marketing automation software supports the APIs that you need in order to efficiently collect your data.

INTEGRATION

Integration can be used as an in-house desktop program or online web sales marketing solution. It can also be part of your company's customer relationship management programs. You should search for a solution that will give you the most flexibility and value for your company's needs.

If you use a common sales management system, look for a marketing automation software program that will integrate with your specific system. Most marketing automation software programs have compatibility with other essential marketing programs. The marketing automation program can be your first step toward having a comprehensive automated system to manage your business.

SUPPORT

Ongoing support is critical for being able to use marketing automation software effectively. The first area of support that you should look for is in set up and implementation, which we have already covered. Your marketing automation software company should help you get up and running with their interface so you can get the most out of their software. In addition to getting up and running, they should provide training and support so you know exactly how to use their program. Some companies may offer formal training programs while others rely on videos and tutorials.

The support should also extend to any technical problems that you may run into while using the software. Requests for help should be answered quickly and you should be able to reach a company representative by phone or email in order to get the help that you need.

Some marketing automation software companies offer community support where users can join together in a forum atmosphere to support one another with issues they may be experiencing. This environment can be helpful and add to the quality you get from the software.

Templates can also be helpful in using the software effectively. Look for a marketing automation software program that comes with pre-written templates you can use for marketing messages. Templates will help you get to the results you want to achieve more quickly, and will assist you with reducing campaign trial and error.

Case Study: Rare Space

Support: Net-Results offers an interactive live demo for prospective clients as well as training webinars for the general public. Customers receive a client login that will take them directly to the client area where there is additional instruction and support.

RED FLAGS FOR MARKETING AUTOMATION SOFTWARE

+ **Limited list of tracking features** – Marketing automation software is a big investment, so you need to have a comprehensive list of ways that you can use the software. If your business is not big enough to use all of the features available, you should still consider using a program that has more than you need. It's better to have more than you need than outgrow a program within a year or two that you'll have to replace.

- **Limited support** – Support is essential for learning any new software or automated system, and this is especially true as you begin to use marketing automation software that will be critical in your company's growth and future profits. Whether it's through the form of training, a support desk or a community forum, the marketing automation software program you choose should have various methods of support.

- **Low quality lead scoring** – Lead scoring is the foundation of the marketing automation software industry. A helpful marketing automation software program will categorize and rank your leads so you can effectively market to them and build your customer base. The more quantifiers that the program uses to rank your leads, the more accurate those rankings will be.

- **Limited integration** – The right choice of marketing automation software for your company will be the program that offers integration with the software and systems that you are already using. While there will be some degree of adjustment while you begin to use new marketing automation software programs you shouldn't expect to have to rebuild your sales funnel or your lead generation process from the ground up.

CASE STUDY RESULTS

Marketing automation is unique for each company that is using it; however, it can be helpful to look at the results that one company secured through using marketing automation software. The following results for Rare Space show the value of a marketing automation solution.

NET-RESULTS CASE STUDY FOR RARE SPACE

As a result of the integration of their current databases with the Net-Results system, Rare Space was able to get more insight into what parts of their site were working and the level of interest of their prospects. One of their newest prospects commented that they assumed that Rare Space was a large firm because of the understanding they were able to offer due to the customization options with marketing automation. Rare Space has realized how important email marketing can be and are making better use of their current clients' data to determine their purchase timeline.

CONCLUSION

The right type of marketing automation software can save you time and increase your conversions. By utilizing the criteria in this chapter, you'll be able to determine if the software has the features and the integration that you need in order to make the right decision for your marketing automation software need.

CHAPTER TWENTY-ONE:

PPC BID MANAGEMENT SOFTWARE

Managing pay per click bidding for your online advertising campaigns can be a time consuming process. Considering the fact that many firms have full time PPC bid management positions, it may be difficult to compete in the industry if you don't quite have the staff or the time to devote to this aspect of online management.

Constant monitoring and updating of pay per click bids is a large task to manage while running a business. However, without constant vigilance your pay per click marketing budget can outgrow your income from your advertising campaign. In order to bridge this gap, you can use PPC bid management software in order to make it easier for you to manage your clicks, run your ads and track your success. You can save yourself time and money by using a reputable pay per click software that will make using PPC marketing all that much more easy and profitable.

PPC bid management tools will help automate bids so that you can take control of your ROI without having to be hands on with PPC management 24/7. PPC bid management software will also help:

o   Increase clicks

o   Improve ad position

o   Decrease costs per click

o   Save time and money

o   Reduce management cost

o   Improve ad campaign performance

In the following chapter, we'll look over the basics of automated PPC bid management software and what you need to consider when you are selecting software.

THE BASICS OF PPC BID MANAGEMENT SOFTWARE

A PPC bid management software tool makes managing PPC campaigns easy by automating the bidding process. If you have a large number of campaigns running that are spread across various demographics and geo targeted areas, automation will help you maintain some control over the situation. PPC bid management tools will give you a bird's eye view of your PPC accounts so you can maintain better control of your spending and your results.

A tool, like the ones evaluated at topseos.com (http://www.topseos.com/rankings-of-best-pay-per-click-bid-management-software), will allow your company to manage multiple campaigns across multiple search engines. You will find that most software will have the same basic function – analyzing bid price and keeping track of the search engines which are displaying your ads. Basic features will also include automatic update bidding, bid gap placement, bid gap squeezes and other features.

Functions of a PPC bid management tool fall into two basic categories:

o   Bid management – Bid management should include several bidding strategies that you can set up that will automatically go into effect on your accounts. The strategies can be applied to groups of keywords or to individual keywords. The rules should be applied consistently to multiple search engines in order to ensure consistency. Good pay per click management software should allow you to change the unit that you are bidding on. It will allow you to set a certain cost per click, bid a certain amount per conversion and track your results so you can more effectively use the advertising model.

o Reporting – Reporting will allow you to understand what is going on with your PPC accounts and how you can improve your results. A PPC bid management tool should allow you to track ROI, view conversions and understand how individual keywords impact your advertising.

Case Study: Find Me Faster

PPC Bid Management Company: Acquisio SEARCH (http://www.acquisio.com/)

Find Me Faster, a Search Marketing Agency based in Nashua, NH, services clients throughout the US and Canada, and specializes in strategic development and implementation of paid search for online and offline marketers.

Tracking and reporting performance across multiple search platforms has always been a tedious, time consuming, manual process for Account Managers at Find Me Faster. The time spent collecting and organizing paid search data was eating away at time they could have spent on strategy and client service, leaving them struggling to meet deadlines and urgent requests.

In addition to these basic features, the following characteristics should be found in any professional bid management software.

USER INTERFACE

The user interface of the PPC bid management tool is essential in your success with working with the tool. If the user interface is simple and easy to understand, you'll be more likely to use the tool and reap the benefits. You'll easily be able to train your team members to use the software as well. The user interface can include the visual design of the program, the intuitive nature of the controls and the various reports that can be generated from the program.

Look for a PPC bid management tool that will incorporate several different functions into one simple interface. You should be able to optimize each campaign individually to maximize your ROI. You should be able to see a top level view of the accounts, the campaigns and ad groups and then track trends of all of those aspects of the accounts.

Be sure that the bid management tool will allow you to increase the bids from the tool quickly and easily. In the world of pay per click a matter of a few minutes can make a difference in how effective your campaigns are. Look at screen shots of the user interface and read a full list of the capabilities before you invest in a PPC bid management tool.

Case Study: Find Me Faster

User Interface: Although there weren't any specific references to User Interface in the Find Me Faster case study. Acquisio Search is a web-based client that offers a bird's eye view of multiple accounts. There is one dashboard that allows the users to access multiple accounts for multiple clients. It uses graphical elements to convey important information – like how the ad click-throughs are increasing or decreasing.

Case Study: Find Me Faster

Features: One of the most beneficial features of the Acquisio SEARCH software, from Find Me Faster's perspective, is the ability to track multiple campaigns across multiple search platforms. Because Find Me Faster handles pay per click advertising for a variety of clients, it's important that they are able to collect and organized paid search data. Automating reports allowed them to send out the reports on the first day of every month.

FEATURES

The features of the PPC bid management tool will help you manage your accounts. A PPC management software tool can help you accomplish a variety of different tasks dealing with PPC management. PPC management software should, at the very least, include the following features for you to be able to use it effectively.

o   Automating campaign and keyword bidding.

o   Ability to achieve the target position and reduce CPC costs.

o   Add, edit and delete campaigns, ads and keywords.

o   Manage a variety of different search engines at once.

o   Create and import ad campaigns across search engine platforms.

o   Schedule and manage ad campaign reports.

o   Create performance reports across a variety of metrics.

o   Increase or decrease budgets.

o   Start and pause campaigns.

o   Review bidding history and performance.

In addition to these features, look for the ability to customize your experience by calling up specific reports and editing the content of these reports. The features included in the PPC bid management tool should provide a wide variety of customizations so you can take in, review and customize your data and your bid strategy.

Case Study: Find Me Faster

Effectiveness: Find Me Faster noted that the reports generated by Acquisio SEARCH were effective, professional and able to be customized. They were able to consolidate their month-end reporting cycle to just a day with the help of the software.

EFFECTIVENESS

The effectiveness of a PPC bid management tool can be seen in the satisfaction of the existing customers and your own personal experience.

The best way to determine the effectiveness of a PPC bid management tool is to research customer feedback thoroughly and engage in a trial offer.

Most PPC bid management companies will offer free trials of their service so you can determine their effectiveness. During your free trial you need to run the service through a variety of different tests to see how it will deal with a variety of different tasks. Run the program through its paces to see how effective it will be in real time so you know it will be reliable. Start a handful of PPC campaigns and test out the management and reporting capabilities. You want to make sure that the program will work as it promises before you invest your time and put the financial security of your advertising in the hands of a PPC bid management program.

Case Study: Find Me Faster

Support: Acquisio SEARCH offers full contact information, online tutorials and ongoing support for their customers which makes it possible for them to get the most out of the system quickly.

SUPPORT

Getting support from a PPC bid management company is essential to being able to work with the program effectively. From live support over the phone, email support while you are running the program or video tutorials to help you learn the ins and outs of using the program, support should be a major part of any PPC bid management company's service.

Besides customer training, a PPC bid management company can also offer help in developing a marketing strategy. PPC advertising is a complex model and in order to master it you should be able to rely on the support of the PPC bid management company. Look for a level of support that you can feel comfortable with. It's important that you find a company that offers you enough support so that you can get started with the system and make it a relied upon part of your business.

STABILITY

Working with a PPC bid management tool means that you'll have to rely on them month in and month out in order to be able to use it effectively. It's important to choose a company that has a long track record in the industry. Seek out companies that actively offer support, customer interaction and that are involved in the industry. Give preference to a company that maintains a blog, has a high rating with the Better Business Bureau and has little to no bad reviews online.

Case Study: Find Me Faster

Stability: Find Me Faster relies upon the stability of the software to create their customer reports each month. Since the software is web based, Find Me Faster and other clients can be assured that the data will be secure and accessible month in and month out.

RED FLAGS FOR PPC BID MANAGEMENT

- **Complicated design and clunky interface** – You are less likely to use a program if you can't work with it efficiently or you don't understand how to use it. A trial offer should give you enough hands on experience to be able to tell if you can use the program.

- **Limited management tools** – A PPC bid management program should give you the capabilities to manage all of your PPC accounts across many different platforms. Having a program that only offers coverage on Google or Yahoo won't allow your PPC strategy to grow. You should also avoid a program that doesn't update your statistics in real time or doesn't let you change bids and prices from within the tool.

- **Lack of personalized support** – A user manual can help you get started with a program but you should look for a company that goes the extra mile and offers live support that you can use in order to get started. Training programs and teleseminars are also a good sign.

CASE STUDY RESULTS

Looking at the case study in this chapter shows how PPC bid management software can help automate processes that previously took up dozens of man hours that could be better spent elsewhere.

ACQUISIO SEARCH'S CASE STUDY FOR FIND ME FASTER

Find Me Faster was pleased with the results of their experience with Acquisio SEARCH. Their time spent at the end of the month creating reports for clients was reduced from a week to just a day. This allowed them to invoice more quickly and also meet the needs of clients who needed reports on demand. The white label nature of Acquisio SEARCH's PPC management software allowed them to brand the reports and customize them to meet the needs of their customers.

CONCLUSION

Finding reliable PPC bid management software can greatly reduce the amount of time and effort this is spent on handling pay per click management. With the right software, you'll be able to make your campaigns more profitable and analyze the results more completely. Whether you are handling your own PPC campaigns or need a tool that will assist you with managing the PPC needs of your clients, using the previously listed criteria will help you select a reliable software program.

CHAPTER TWENTY-TWO:

SEO SHOPPING CART SOFTWARE

S EO shopping cart programs will give your company a framework that will be searchable by both your website visitors and in the search engines as well. Shopping cart programs will automate a lot of the processes that your visitors need to complete in order to successfully make a purchase from your website.

At a very basic level a shopping cart program is a simple database that organizes the items you sell on your website. From a programmer's point of view, it's very easy to create a shopping cart program that will deliver the right results for website visitors. However, when it comes to being search engine friendly, many shopping cart programs leave a lot to be desired.

In this chapter, we'll be going over the major qualities you need to look for in an SEO shopping cart program provider. These qualities will make your website functional from a search engine optimization point of view as well as create an enjoyable experience for your website visitors.

ESSENTIAL FEATURES OF AN SEO SHOPPING CART PROGRAM

SEO is vitally important to an online store in order to get their products exposed in the right search engine results pages. The goal is to get your "blue widget" page to show up in the first page of results for "blue widgets." First generation shopping carts can help your website visitors navigate through your site and purchase the products they are interested in, but an SEO shopping cart takes the functionality of a shopping cart program one step further.

Most popular shopping cart programs are designed by back end programmers who are primarily concerned with the functionality of the shopping cart – successful searching and purchases. However, this leaves a large part of the success of a website – search engine optimization – off of the table. SEO shopping carts bridge that gap between shopping cart functionality and search engine optimization.

SEO shopping cart programs may seem like foreign territory if you aren't experienced with the technical side of running a business. With the following guidelines you won't be able to accurately evaluate an SEO shopping cart program to see if it will meet your needs and assist your company with creating a better online presence. For further assistance, see the listings at topseos.com located at http://www.topseos.com/rankings-of-best-seo-shopping-cart-software.

Case Study: Heater-Store.com

SEO Shopping Cart Company – SearchFit (http://www.searchfit.com/)

Heater-Store.com's goal is to become the premier online source of heating products. They offer quality, name brand products for both commercial and residential use. Their inventory includes a broad range of products from small office heaters to the large 1.5 million BTU industrial heaters used in the coldest parts of the world. Heater-Store.com's strategy is to gain a competitive edge in the market by:

1. Offering a broad range of high quality products at the lowest possible prices and outstanding customer service.

2. Maximize conversion rates by offering the best possible shopping experience.

3. Minimize the cost of driving traffic to their website.

Although Heater-store.com had been powered by SearchFit for several years, the most recent version of the site was updated with a new design and utilized several of SearchFit's new features.

GRAPHICAL USER INTERFACE (GUI) FEATURES

GUI features refer to the visual aspects of the SEO shopping cart program that you'll encounter as a backend user. Once your SEO shopping cart company installs the program and gets it up and running, you should be familiar enough with backend interface so that you can add products and make adjustments as you see fit.

The GUI features of an SEO shopping cart program will make your updates of the program easy and manageable. There's no point in investing in a shopping cart program and having to call for help at every step of the way when you need something minor changed. Look for an SEO shopping cart program that makes updates and changes simple for you as the backend user. If you can test out the system before purchasing or see detailed screenshots of the backend interface before you make the investment it will help you decide whether or not the GUI features are going to be acceptable and to your liking.

In addition to the backend user GUI, you should also evaluate the SEO company in terms of the graphical user interface that your users will be encountering. The shopping cart program should seamlessly fit in with the rest of your design and site functionality.

Case Study: Heater-store.com

GUI Features: One of the key GUI features used by Heater-store.com was SearchFit's SEO Naming Template. This interface allowed Heater-store.com to increase the presence of important keywords on the website and create new product groups with just a few clicks. This contributed to Heater-store.com's overall improvement in results with regards to search engine results placement.

FEATURES

The features in an SEO shopping cart program will help you determine whether or not it will be able to meet your needs. A professional SEO shopping cart program should feature a few basic principles which should be automated to make your website more SEO friendly. In addition to the basic principles of automating the shopping process, your SEO shopping cart program should allow you to incorporate a variety of features that will help your users have a positive experience on your site. The platform of your website should incorporate a variety of different options to make your website fully functional.

Look for the following features in your SEO shopping cart program to be sure it will create a valuable user experience for your visitors.

SALES AND SPECIAL DISCOUNT FEATURES:

o   Integration with sales software like QuickBooks

o   Discount coupons

o   Affiliate programs

o   Gift certificates

SHIPPING AND TAX ASSISTANCE:

o   Automatic postage calculations (with options for UPS, FedEx, USPS, etc)

o   International postage calculations

o   Automated calculation of state and federal taxes

PAYMENT OPTIONS:

o   Multiple payment options

o   Real-time credit card processing

o   Multiple currency displays

o   Recurring billing capabilities

o   Electronic gateway options

o   Bulk purchase discounts

CUSTOMER SERVICE OPTIONS:

o   Order taking options (website, over the phone, fax)

o   Recognition of repeat customers

o   Automatic confirmation emails

o   Automated lost password recovery

o   Site search navigation and security:

o   In-site search engine

Look for an SEO shopping cart program that has features that are more than what you presently need. Although you may not need specific features now, you'll want to be sure that your shopping cart program can grow with your business as you need more features in the future.

Case Study: Heater-store.com

Features: Heater-store.com made use of many of SearchFit's features that are designed to help ecommerce stores become more search engine optimized. Heater-store's website update included:

Showcasing products with product groups

Improving layout and structure of top level category pages

Enhancing category pages with new shopping features

Enhancing detailed product displays with "add to wish list" and "email a friend" functions.

Adding a product filter navigation system

Adding a comparison shopping engine geed system

Adding a advanced website search function

CUSTOMIZATION

Customization is essential for an SEO shopping cart program because your website will be unique to your products and services. It's absolutely essential that you get a high level of customization with your SEO shopping cart provider. This way your shopping cart program can reflect the exact nature of your business.

One of the major areas of customization is the look and feel of the shopping cart program. The shopping cart area of your website should blend seamlessly with the rest of your website. The individual pages should be able to be customized so that your products are displayed in the way that is most conducive to your sales process.

The shopping cart design should have all of the popular fields like META titles, page titles and META keywords. These fields will be customized based on your products and your keywords so each page can be indexed properly by the search engines. You should be able to make changes to these areas as you see fit in the future. Look for customization aspects like the ability to add additional information like "special price."

A good shopping cart design will allow you to customize the interface to meet your company's needs and create a better user experience for your customer, while still keeping some basic tenants of shopping cart interaction intact so customers are familiar with the layout.

Case Study: Heater-store.com

Customization: Heater-store.com was able to customize a great deal of their shopping cart interface. Not only were they able to improve on keyword selection for their category pages and product pages, but they were also able to structure the top level categories so that their best products were displayed. They were able to utilize the "on sale," "best sellers" and "new items" categories to highlight important products that would be of interest to buyers.

Case Study: Heater-store.com

SEO Friendly: SearchFit offers many options to increase search engine optimization and improve organic search engine traffic. The SEO Naming rules feature helped Heater-store.com automate the process of creating optimized product page names, titles, keywords and meta titles, descriptions and tags.

SEO FRIENDLY

The most important aspect of an SEO shopping cart program is the search engine optimization aspects. Keywords should be placed in key locations so that your pages are indexed correctly. The following areas should be covered by your SEO shopping cart program.

KEYWORD RICH URLS

URLs are one of the most important areas in order to achieve better search engine optimization. Your product pages are essential in your site architecture from an SEO point of view and a user point of view. It is absolutely vital that these pages are optimized for your keywords. Most shopping carts will place your pages in the following structure:

http://www.samplepage.com/department-name/product-name

If your SEO program defaults into this format, see if you can change it to one of the following, more optimized, formats.

http://www.samplepage.com/product/product-name/
http://www.samplepage.com/prod/product-name/
http://www.samplepage.com/p/product-name/

Department names aren't really necessary because sometimes products may fit under many different categories. It's more important for you to use the product name as much as possible for optimization purposes. When website browsers are looking for a specific product they are unlikely to search for "dinnerware" or "table setting." They'll be searching for more specific keywords like "blue plates" or "red oval tablecloth."

You can use categories like "dinner ware" or "specials" in order to organize your site but they don't need to be part of the URL display structure. Your SEO shopping cart program should either use a structure for URLs like the ones in the example or allow you to edit the URL structure so that it is SEO friendly.

Product names should be descriptive and not alpha-numeric unless there is a reason that they absolutely have to be. Be sure that your SEO shopping cart program will keep your product URLs at a limit of 3 to 5 words for the product names. This will ensure the URLs are manageable for search engine spiders as well as your users.

Department and category pages are some of the most contested areas of SEO shopping cart design. Experts disagree on whether having department or category pages are really of value to your overall site's design and optimization. If there is no editorial content on the category pages, the page doesn't need to be indexed. If you opt to have department and category pages listed on your site, your SEO shopping cart company should offer you the option of leaving these pages out of the indexing process.

TITLE, DESCRIPTION AND KEYWORD META DATA

SEO cart programs should allow you to have complete control over the Title, META Description and Keyword META data so that you can optimize your pages. Relevant keywords need to be included in these important areas and your SEO shopping cart program should make their integration easy and seamless. You should also have the option of placing copy into the META description of the website in order to boost the chance of getting traffic from search engine results. Your SEO shopping cart system should allow you to customize the following areas.

H1 HEADER TAGS

H1 header tags will increase the odds that a page will rank well for a keyword term that is included within the product tags. Ideally, an SEO shopping cart program will draw keywords for your header tags from your product detail screens.

ALT TAGS

Alt tags are the text that is displayed when a search engine spider cannot read images. These can be used by search engine spiders to classify and understand the image content. Use descriptive, keyword friendly alt tags in order to give your shopping cart more SEO power.

CANONICAL ELEMENTS

Canonical elements are tags that are put on a page in order to ensure that the search engine is indexing the right version of the URL. This minimizes confusion for the search engine spiders and can optimize your pages for your desired keywords and product descriptions. Your SEO shopping cart program should make this a natural part of your pages.

OPTIMIZED ANCHOR TEXT

Your SEO shopping cart program should automatically make your links keyword rich for extra optimization. The links should point to category and product pages in your catalog so that they are optimized for descriptive keywords that relate to the category or the product page.

OPTIMIZED LINK STRUCTURE

The SEO shopping cart program should arrange your product and category pages in such a way that the search engine spiders can logically crawl through your site to index it appropriately. It should also create a structure that is accessible and easily reached by consumers. Putting products into category structures that keep your site organized and optimized are key to helping both search engines and visitors understand your site and navigate it successfully.

AUTO-GENERATED ROBOTS.TXT AND XML SITEMAPS

Sitemaps, are of extreme importance in terms of search engine optimization. SEO shopping cart programs should automatically generate site maps that will be submitted to search engines in order for the search engines to index your site's pages. In addition, a robot.txt file will be generated in order to point visiting search engine spiders toward your site map within your domain.

For more on the details of search engine optimization, see the SEO chapter of this book. The levels of optimization that your SEO shopping cart company can employ are very deep and look to that section of this book for a more complete understanding.

Case Study: Heater-stores.com

Stability: Heater-stores.com had been a SearchFit client previously before their newest site update, so they were familiar with their history and comfortable continuing to use them.

STABILITY

Stability is an important factor in whether or not you can rely on a company for years to come. If the SEO company has been in business for a long time, it's a good sign. The SEO shopping cart company should have a strong track record of success with other customers so you can be sure they'll be there for you.

Look for a company that has a strong presence in the industry and that maintains active contributions – such as maintaining a blog, holding ongoing training, speaking at conferences, etc. A simple online search of the potential company name should pull up a track record that will give you a sense of the reputation of the company and whether they'll be around for the long term.

RED FLAGS FOR SEO SHOPPING CART PROGRAMS

+ **Lack of customization** – Having a customized shopping cart program, whether it's through the design, the product display or the overall site structure, is essential in making your website a unique experience for the user. Your customization options should range from small tweaks to large changes so you can get the exact design that you want.

+ **Lack of true SEO knowledge** – There's a big difference between a shopping cart that works and a shopping cart that delivers optimized pages that will help you rank well in the search engines. Based on the categories outlined in this chapter and the SEO chapter of this book, you'll be armed with the basic information you need to determine if a company understands the basic concepts of SEO and has integrated them into their program.

+ **Poor GUI features** – GUI is essential in being able to make changes and essential to allowing your visitors to use the site successfully. Evaluate screen shots of the GUI and see if you can understand the interface.

+ **Lack of installation support** – Installation is critical to having an SEO shopping cart that works effectively. If the program is not installed, your website won't get the SEO benefits and definitely won't have the functionality that you're looking for. See if the company offers ongoing support during and after installation.

CASE STUDY RESULTS

The case study in this chapter helps translate the abstract qualities of an ideal SEO shopping cart solution into a real world example. By looking at the results of the improvements that SearchFit made with Heater-store. com's website, we can see how effective an SEO shopping cart can be.

SEARCHFIT CASE STUDY FOR HEATER-STORE.COM

Heater-store.com was already using SearchFit as their SEO shopping cart provider when they decided to make a few improvements with their system. Their increased usage of SearchFit's existing features and incorporation of new features resulted in a 105% increase in ecommerce conversion rates. They also saw a 62% increase in average order value and a 5% decrease in bounce rate. These results were achieved in a two month period.

CONCLUSION

Setting up an ecommerce store with an easy to navigate system that also complies with the needs of search engines can be a complicated task without an SEO shopping cart provider. The qualities outlined above will put you one step closer to finding a provider that can help you automate product orders and attract organic traffic from the search engines as well. Looking at the results achieved by Heater-store.com with the SearchFit system, it's clear to see that SEO shopping cart programs are worth the investment.

BIBLIOGRAPHY

Allem, Mae. "14 Smart Questions to Ask Conversion Optimization Consultants." *Invesp Blog – Conversion Rate Optimization Blog.* July 7, 2008. (accessed February 2, 2010). http://www.invesp.com/blog/conversion-optimization/14-smart-questions-to-ask-conversion-optimization-consultants.html

Ash, Tim. "Emotional Motivators in Landing Page Optimization." Web Marketing Today. September 22, 2009. (accessed January 22, 2010) http://www.wilsonweb.com/conversion/ash-emotional-motivators.htm

Ash, Tim. (2008) Landing Page Optimization: The Definitive Guide to Testing and Tuning for Conversions. Indianapolis, Indiana. Wiley Publishing.

Blanchard Olivier. "Basics of Social Media Roi." *Hosted on SlideShare.* August 24, 2009. (accessed December 29, 2009) http://www.slideshare.net/thebrandbuilder/olivier-blanchard-basics-of-social-media-roi

Bowman, Jessica. "SEO Site Audit: A Wise Investment For All Companies." March 5, 2008. (accessed March 27, 2010) http://searchengineland.com/seo-site-audit-a-wise-investment-for-all-companies-13511

Brinker, Scott. "A Completely Different Kind of Landing Page Optimization." *Search Engine Land*. October 22, 2008. (accessed January 21, 2010) http://searchengineland.com/a-completely-different-kind-of-landing-page-optimization-15201

Brogan, Chris. "What I Want a Social Media Expert to Know." *ChrisBrogan. com*. April 15, 2008. (accessed December 28, 2009) http://www.chrisbrogan.com/what-i-want-a-social-media-expert-to-know/

Clark, Brian. "Great Landing Pages Turn Traffic Into Money." *Copyblogger*. (accessed January 27, 2010). http://www.copyblogger.com/landing-pages/

Clay, Bruce. "Search Engine Optimization (SEO)." *Bruce Clay, Inc.* (accessed October 10, 2009) http://www.bruceclay.com/web_rank.htm

Clay, Bruce. "Pay Per Click (PPC) Methodology." *Bruce Clay, Inc.* (accessed November 12, 2009) http://www.bruceclay.com/web_ppc.htm

Clifford, Stephanie. "Video Prank at Domino's Taints Brand." *The New York Times – Media and Advertising*. April 15, 2009. (accessed December 31, 2009) http://www.nytimes.com/2009/04/16/business/media/16dominos.html

Davies, Dave. "What to Look for in An SEO." *Beanstalk Search Engine Optimization*.(accessed September 12, 2009) http://www.beanstalk-inc.com/articles/seo/in-an-seo.htm

"Does your company need to hire an SEO firm?" *Yesup SEO*. (accessed September 30, 2009). http://www.yesupseo.com/search-engine-optimization-seo-providers.html

"E-commerce Feature: Storefront Setup." *SEO-Cart*. (accessed March 2, 2010) http://www.seo-cart.com/shopping-cart/e-commerce-features-storefront-setup

Elliance. "Search Illustrated: Search Engine Click-Thru Behavior; You've Got to Be In the Top Ten!" Search Engine Land. August 7, 2007. (accessed September 13, 2009) http://searchengineland.com/search-illustrated-search-engine-click-thru-behavior-youve-got-to-be-in-the-top-ten-11883

Enge, Eric. "Content is King, Baby!" *SEOMoz*. April 16, 2008. (accessed March 10, 2010) http://www.seomoz.org/blog/content-is-king-baby

Evans, Sarah. "10 Ways to Make Press Releases More SEO Friendly." *Mashable*. November 4, 2008. (accessed February 10, 2010) http://mashable.com/2008/11/04/how-to-make-press-releases-seo-friendly/

Fallows, Deborah, PhD. Search Engine Users. *Pew Internet Research Center Publications*. Janaury 23, 2005. (accessed September 25, 2009)

Fallows, Deborah, PhD. "Search Soars, Challenging Email as Favorite Internet Activity." *Pew Research Center Publications*. August 6, 2008 (accessed October 2, 2009). http://pewresearch.org/pubs/921/internet-search

Fried, Ina. "Yahoo, Microsoft reach search, ad deal." *CNET*. July 29, 2009. (accessed November 8, 2009) http://news.cnet.com/8301-13860_3-10298303-56.html

Gannes, Liz. "The Great Video SEO Frontier." *Bloomberg BusinessWeek*. February 13, 2009. (accessed February 2, 2010). http://www.businessweek.com/technology/content/feb2009/tc20090212_136831.htm

Geddes, Brad. "Ignoring the Content Network? Think Again to Vastly Improve Conversions." *Search Engine Land*. March 30, 2009. (accessed December 21, 2009) http://searchengineland.com/a-unique-look-into-content-network-organization-to-increase-total-sales-17069

Gray, Michael. "Shopping Cart SEO Tips." *Michael Gray – Graywolf's SEO Blog*. December 2, 2009. (accessed March 2, 2010) http://www.wolf-howl.com/seo/shopping-cart-seo-tips/

Harte, Beth. "Social Media Transparency: How Realistic Is It?" *Social Media Today*. March 2, 2009. (accessed December 28, 2009) http://www.socialmediatoday.com/SMC/77818

"Hidden Text and Links." *Google Webmaster Central*. (accessed October 16, 2009) http://www.google.com/support/webmasters/bin/answer.py?hl=en&answer=66353

"How to find the best Search Engine Friendly SEO Shopping Cart." *Big Oak Inc.* (accessed March 27, 2010) http://www.bigoakinc.com/seo-articles/seo-friendly-shopping-cart.php

Jansen, B.J. and Spink, A. Investigating customer click through behaviour with integrated sponsored and nonsponsored results. *Int. J. Internet Marketing and Advertising*, Vol. 5, Nos. 1/2, pp.74–94. 2009. (accessed November 19, 2009)

Jerum, Greg. How Much of Your Money Will Your PPC Manager Waste this Year?: Essential Questions You Must Ask before Hiring a PPC Manager. *Net Return Marketing.* (accessed November 28, 2009)

Jones, Ron. "PPC Bid Management 101." *SearchEngineWatch.* July 13, 2009. (accessed December 1, 2009) http://searchenginewatch.com/3634372

Karjaluoto, Eric. A Primer in Social Media. March 1, 2008. (accessed January 2, 2010) http://www.smashlab.com/papers/item/p/list6ItemID/v/3

Kelly, Rebecca. "Have Online Press Release Gone the Way of the Dodo?" *SEOMoz.* May 6th, 2008. (accessed January 15, 2010) http://www.seomoz.org/blog/have-online-press-releases-gone-the-way-of-the-dodo

"Keyword Stuffing." *Google Webmaster Central.* (accessed October 16, 2009 http://www.google.com/support/webmasters/bin/answer.py?hl=en&answer=66358

Latenslager, Al. "30 Press Release Ideas." Reprinted on *AOL Small Business* from *Entrpreneur.com.* November 11, 2006. (accessed February 1, 2010). http://smallbusiness.aol.com/grow/marketing/article/_a/30-press-release-ideas/20051213183909990019

Lurie, Ian. "10 questions to evaluate an SEO." *Conversation Marketing.* February 2, 2010. (accessed February 8, 2010) http://www.conversationmarketing.com/2010/02/10-questions-to-evaluate-an-seo.htm

Lurie, Ian, "10 Questions to Evaluate a Social Media 'Expert'." *Conversation Marketing.* July 21, 2009. (accessed February 1, 2010) http://www.

conversationmarketing.com/2009/07/10-questions-for-social-media-experts.htm

MacDonald, Dennis. "What Social Media Adoption Model Are You Following?" *Dennis McDonald's Web Site*. August 15, 2007. (accessed December 28, 2009) http://www.ddmcd.com/managing-technology/what-social-media-adoption-model-are-you-following.html

Mackenzie, Josiah. "23 Questions You Must Ask PPC Management Companies." *Hotel Marketing Strategies*. December 23, 2008. (accessed December 15, 2009) http://www.hotelmarketingstrategies.com/23-questions-for-ppc-agencies/

Martelli, Don. "Developing Social Media Strategies: The Power of Community." *SlideShare*. July, 20, 2009. (accessed December 28, 2009) http://www.slideshare.net/donmartelli/develop-social-media-strategies

Mastaler, Debra. "Help, I'm New I Need Links, What Can I Do?" *Beanstalk Search Engine Optimization*. (accessed October 28, 2009) http://www.beanstalk-inc.com/articles/links/10step-links.htm

Mayfield, Anthony. What is Social Media? *iCrossing*. January 8, 2008. (accessed December 29, 2009) http://www.icrossing.co.uk/fileadmin/uploads/eBooks/What_is_Social_Media_iCrossing_ebook.pdf

McGaffin, Ken and Archie Binnie. Linking Matters: How to Create An Effective Linking Strategy to Promote Your Website. *McGaffin.com*. February 6, 2003. (accessed January 4, 2010)

Milian, Mark. "Burger King sacrifices Facebook app after dispute." *Los Angeles Times – Business section*. January 15, 2009. (accessed December 31, 2009) http://latimesblogs.latimes.com/technology/2009/01/burger-king-fac.html

Mody, Melissa. "Effective Press Releases and Distribution Channels." *Website Magazine*. May 1, 2008. (accessed January 15, 2010) http://www.websitemagazine.com/content/blogs/posts/pages/Effective-Press-Releases-and-Distribution-Channels.aspx

Mordkovich, Igor. "17 Most Common PPC Mistakes Web Marketers Make." *BizMord Online Marketing Blog.* January 3, 2007. (accessed December 1, 2009) http://www.bizmord.com/Blog/archives/200

Odden, Lee. "Press Release SEO Tips." *Online Marketing Blog.* November 27, 2007. (accessed February 10, 2010) http://www.toprankblog.com/2007/11/press-release-seo-tips/

Quenet, Daryl. "Part Three of Ten: Site Structure." *Beanstalk Search Engine Optimization.* November 7, 2004. (accessed September 31, 2009) http://www.beanstalk-inc.com/articles/seo/10step-structure.htm

Rhodes, Matt. "Social media as a crisis management tool." *Fresh Networks.* December 21, 2009. (accessed December 31, 2009) http://blog.freshnetworks.com/2009/12/social-media-as-a-crisis-management-tool/

Robinson, Shane. "How to reduce your PPC costs (Episode 1)." *Affiliate Doctors.* December 22, 2009. (accessed January 6, 2010) http://www.affiliatedoctors.com/how-to-reduce-your-ppc-costs-episode-i/

Robertson, Mark. " "Reel" Video SEO Strategies & Best Practices." *Hosted on SlideShare.* February 10, 2010. (accessed February 12, 2010). http://www.slideshare.net/ReelSEO/reel-video-search-engine-optimization-webinar-by-reelseo

Robertson, Mark. "Video Accessibiltiy, Closed Captions & Video SEO." *ReelSEO.* March 30, 2008. (accessed February 12, 2010). http://www.reelseo.com/video-accessibility-closed-captions-video-seo/

Robertson, Mark. "Video Search Engine Optimization Tips from Truveo." *ReelSEO.* December 11, 2009. (accessed February 12, 2010). http://www.reelseo.com/video-search-engine-optimization-tips-from-truveo/

Robertson, Mark. "Video+SEO – Best Practices for Online Video Publishing & E-commerce." *Hosted on SlideShare.* June 2009. (accessed February 10, 2010). http://www.slideshare.net/ReelSEO/video-seo-best-practices-for-online-video-publishing-ecommerce

Robertson, Mark. "Youtube Keyword Research Tool for Video SEO." *ReelSEO*, June 23, 2008. (accessed February 10, 2010). http://www.reelseo.com/youtube-keyword-research-tool/

"SEO Tutorial Part 9 – Off-Site Optimization Details." InlineSEO. October 22, 2008. (accessed November 10, 2009) http://www.inlineseo.com/blog/2008/10/22/off-site-optimization/

Smith, Kraig. Search Engine Advertising Update – Q309. *AdGooRoo*. October 13, 2009. (accessed November 9, 2009) http://www.adgooroo.com/research/Search%20Engine%20Advertiser%20Analysis%20-%20Q309.pdf

Social Media RFP Template. *SocialMediaGroup.com*. (accessed January 4, 2010)

Snow, Shane. "How Web Video SEO is Finally Coming of Age." *Mashable*. February 1, 2010. (accessed February 2, 2010). http://mashable.com/2010/02/01/web-video-seo/

"Step 2. Web Site Audit." *Stepforth*. (accessed March 27, 2010) http://www.stepforth.com/6-step-plan/web-site-audit/

"The CAN-SPAM Act: A Compliance Guide for Business." *Federal Trade Commission*. September 2009. (accessed March 28, 2010). http://www.ftc.gov/bcp/edu/pubs/business/ecommerce/bus61.shtm

"The Advantage of a Video Spokesman." (accessed March 25, 2010) http://www.web-design-worcestershire.com/tag/virtual-spokesperson/

"There are no magic tricks to good SEO." *Yesup SEO*. (accessed September 30, 2009). http://www.yesupseo.com/magic-tricks-seo.html

Van Grove, Jennifer. "As Whole Foods Boycott Grows on Facebook, Brand Perception Drops." *Mashable – The Social Media Guide*. August 24, 2009. (accessed December 31, 2009) http://mashable.com/2009/08/24/whole-foods-brand-perception/

Wall, Aaron. "Automated PPC Search Engine Bid Management Software." *Search-Marketing.info*. (accessed April 6, 2010) http://www.search-marketing.info/search-engines/price-per-click/autobid-tools.htm

Wall, Aaron. "SEO & PPC Competitive Analysis & Keyword Research Tools." *SEOBook*, July 15, 2005. (accessed November 8, 2009) http://www.seobook.com/archives/001013.shtml

Weintraub, Marty. "SEO Site Audit, a Guerilla Webmaster's Guide." *Aim Clear Blog*, May 18, 2009. (accessed March 27, 2010) http://www.aimclearblog.com/2009/05/18/seo-site-audit-a-guerrilla-webmasters-guide/

"Web Site Audit In-House." *Web CEO*. (accessed March 27, 2010) http://www.web-ceo-site-auditor.com/index.htm

"What does Bounce Rate mean?" *Google Analytics Help*. (accessed January 8, 2010). http://www.google.com/support/googleanalytics/bin/answer.py?hl=en&answer=81986

"When hiring an SEO firm, consider this…" *Yesup SEO*. (accessed September 30, 2009) http://www.yesupseo.com/when-hiring-an-seo-firm-consider-this.html